CLIMATE GHOSTS

The MANDEL LECTURES *in the* HUMANITIES
at BRANDEIS UNIVERSITY

Sponsored by the Jack, Joseph and Morton Mandel Foundation.
Director and chair, Professor Ramie Targoff.

The Mandel Lectures in the Humanities were launched in the fall of 2011 to promote the study of the humanities at Brandeis University, following the 2010 opening of the new Mandel Center for the Humanities. The lectures bring to the Mandel Center each year a prominent scholar who gives a series of three lectures and conducts an informal seminar during his or her stay on campus. The Mandel Lectures are unique in their rotation of disciplines or fields within the humanities and humanistic social sciences: the speakers have ranged from historians to literary critics, from classicists to anthropologists. The published series of books therefore reflects the interdisciplinary mission of the center and the wide range of extraordinary work being done in the humanities today.

For a complete list of books that are available in the series,
visit brandeisuniversitypress.com

Nancy Langston, *Climate Ghosts: Migratory Species*
    in the Anthropocene
David Der-wei Wang, *Why Fiction Matters in Contemporary China*
Wendy Doniger, *The Donigers of Great Neck: A Mythologized Memoir*
Ingrid D. Rowland, *The Divine Spark of Syracuse*
James Wood, *The Nearest Thing to Life*
David Nirenberg, *Aesthetic Theology and Its Enemies:*
    Judaism in Christian Painting, Poetry and Politics

Nancy Langston

CLIMATE GHOSTS

Brandeis University Press | Waltham, Massachusetts

Migratory Species in the Anthropocene

Brandeis University Press
© 2021 by Nancy Langston
All rights reserved
Manufactured in the United States of America
Designed by Mindy Basinger Hill
Typeset in Adobe Caslon Pro and ITC Avant Garde Gothic Pro.

For permission to reproduce any of the material in this
book, contact Brandeis University Press, 415 South Street,
Waltham MA 02453, or visit brandeisuniversitypress.com

Library of Congress Cataloging-in-Publication Data

Names: Langston, Nancy, author.

Title: Climate ghosts: migratory species in the anthropocene /
Nancy Langston.

Description: Waltham, Massachusetts: Brandeis University Press, [2021] |
Series: The Mandel lectures in the humanities at Brandeis University |
Includes bibliographical references and index. | Summary: "Langston focuses
on three ghost species in the Great Lakes watershed—woodland caribou,
common loons, and lake sturgeon. Their traces are still present in DNA,
small fragmented populations, or in lone individuals. We can still
restore them, if we make the hard choices necessary for them to survive"—
Provided by publisher.

Identifiers: LCCN 2021020966 (print) | LCCN 2021020967 (ebook) |
ISBN 9781684580644 (cloth) | ISBN 9781684580651 (paperback) |
ISBN 9781684580668 (ebook)

Subjects: LCSH: Endangered species—Great Lakes Region (North
America) | Endangered species—Great Lakes Region (North America)—
Conservation. | Migratory animals—Climatic factors—Great Lakes
Region (North America)

Classification: LCC QL84.22.G74 L36 2021 (print) | LCC QL84.22.G74 (ebook) |
DDC 333.95/420977—dc23

LC record available at https://lccn.loc.gov/2021020966
LC ebook record available at https://lccn.loc.gov/2021020967

5 4 3 2 1

CONTENTS

ILLUSTRATIONS

PLATES

Color plates appear after page 100.

n a crisp fall day, my environmental studies students quick-ened their pace through the coastal foredunes of Lake Michigan, in Sleeping Bear Dunes National Lakeshore. Endangered pitcher thistle, drought-tolerant juniper, jagged wind-sculpted jack pine, glimmer of an interdunal pond, talk of ghost forests and even long-ago ghost towns buried by living sand dune complexes that migrated with westerly winds; the students endured all these unfamiliarities with nods and smiles. So patient with us elders as we constantly stopped to talk and point, they longed to crest the final hill, to descend into that last swale, before escaping to the opal freshwater sea that awaited them with open arms of winds and waves.

But that day, another scene awaited. Washed up on the beach were hundreds of dead migratory waterbirds, their bodies deformed like partially deflated birthday balloons. The regal necks of common loons—black-feathered with green shimmers, and ringed with a white necklace befitting their cosmologic lineage as clan chiefs—were now contorted as a nightmare. Hollow darkness had snuffed the royal red glow from their eyes. It wasn't just that so many died—over 3,000

FIGURE 0.1 University students on a trip to Sleeping Bear Dunes National Lakeshore. *L. Heasley*

birds at Sleeping Bear alone—or that they died terrified as botulism toxins paralyzed their muscles until they suffocated and drowned. A crushing indignity was also visited upon these birds. We visitors were not equipped to perform the rites of respect and love that these avian relatives of ours deserved on their death-beach. No matter how heartfelt, our temporary tourists' stares were a sacrilege. Our averted gazes as we walked were also a sacrilege. How does one apologize, or make amends?

Loon pairs form some of my earliest memories of two little lakes in northwestern Wisconsin. My grandparents—poor, hardscrabble farmers working poor, sandy land—owned modest properties near their farm, backwoods "forties" on Loon Lake and Bass Lake. There, Grandma in her polyester dress and Grandpa in his roomy bib overalls would take the occasional sojourn. These were too brief to call vacations, or even staycations; they were more like long picnics. My grandma loved the generations of loons who called to her from the lakes and all the other life around her. She was a naturalist at heart. When the majestic century-old white pines and oaks that shaded her shabby farmhouse crashed to the ground in a wind shear, she couldn't call her home "home" anymore, and she moved to a low-income retirement apartment in town. My grandma grieved those trees.

Standing on Loon Lake, my family didn't talk about the Native peoples who *still* live in rural Wisconsin (and also Michigan, Minnesota, and Ontario), and who for time immemorial have honored, protected, and been nurtured by consent of the common loon, lake sturgeon, and other powerful kin. I never learned about the capacious love of a whole people for their nonhuman world, or their capacious grief for wounds to that world. Though I didn't know it then, my grandma and I were more than transient property owners. We were occupiers of sacred ancestral lands. Our hearts thrilled to the tremolo of the loons, but we were of the dominant settler culture. The loon's song was fading beneath our suffocating weight. Its echoes would haunt a Lake Michigan beach.

Three years after the bird kills on Lake Michigan, I walked at sunset along another Great Lakes beach with my friend Nancy Langston and her two dogs, Juneau and Tiva. Tiva was a pit bull mix, and I'm leery of pit bulls. Nancy patiently explained the breed's evolutionary and cultural history, and how they were once nanny-dogs to children of Irish working-class immigrants in America. (Later in Nancy's one-room cabin, which perched on a bluff overlooking Lake Superior, Tiva exemplified her own history by collapsing limp across my lap, gazing up at her new ward.) Meanwhile no dead loons apparated on our radiant ribbon of sand as the exuberant dog grabbed ever-larger limbs of driftwood with her mighty jaws. But their ghosts were with us.

Nancy was soon to leave a faculty position at the University of Wisconsin that colleagues considered a career pinnacle. She would join the new Great Lakes Research Center at Michigan Technological University, in the remote Keweenaw Peninsula of Lake Superior. Nancy and her husband Frank Goodman flattened the aspirational "pinnacle" so that it could include duty to an inland sea. On this

FIGURE 0.2 Nancy Langston and Tiva on a Lake Superior beach.
L. Heasley

finger of land extending into Superior, Nancy's research and home
were now *in place.*

For many of us who work in the lands and waters of environmental
history, Nancy's scholarship has been both formative and formidable.
Her first book was *Forest Dreams, Forest Nightmares: The Paradox of
Old Growth in the Inland West.* It's hard to convey the revelation of
Forest Dreams in 1995, because today we take its pioneering insights
as a given. Rather than approaching old-growth forests in the Blue
Mountains of Oregon from the perpetual rhetoric of economy vs. environment, Nancy started with science. What did early forest scientists
know, ecologically speaking, about these western forests—Douglas fir,
grand fir, western ponderosa pine? How did federal foresters act on
that early science to manage public lands?

The paradox turned out to be this: there was once an alternative
future, based in science, that could have forestalled forest tragedies to

come. In the early twentieth century, scientific studies were in place that could have assured a sustainable supply of timber, while also sustaining a healthy forest. Yet each new decade further depleted the timber supply and enfeebled once-magnificent forests through a combination of overharvesting, insect ravages, and catastrophic fires. Why?

Why did forest science fail the Blue Mountains and other forests in the American West? Notwithstanding the economic and political pressures on foresters everywhere, Nancy's research revealed that in the semi-arid Blues, they made decisions based on the ecology of completely different forests—forests in the humid eastern United States, where these foresters were educated. Foresters managed the Blue Mountains in Oregon as they might have the Porcupine Mountains in Michigan. This was not a failure of science at all; it was a failure to *see* the actual forest. For far too long, professional forestry was *out of place.*

In another book, Nancy examined the early science preceding a different disaster. This time the study site wasn't a forest; it was that most disputed of places, the female body. For *Toxic Bodies: Hormone Disruptors and the Legacy of DES,* she trained a historian's metaphorical hysteroscope on the research of DES, diethylstilbestrol, a synthetic hormone invented in 1938 for women "suffering" from menopause. In 1947, the Food and Drug Administration also approved DES to prevent miscarriages. From then until the early 1980s, doctors submitted to the will of pharmaceutical companies, and prescribed DES to millions of pregnant women.

The problem? Well, for one, a 1953 study showed DES did not prevent miscarriage. Indeed, women injected with DES had higher rates of miscarriage. It gets worse, said Nancy. DES became a routine protocol despite evidence that it caused cancer, disrupted human and animal reproduction, and forever changed the lives of girls born to mothers injected with the toxin. These DES daughters would experience painful, uncertain pregnancies, even infertility, as well as an unnatural risk of vaginal cancer. Each injection of DES had been a callous act of injustice. Collectively they were an atrocity against generations of mothers and their daughters. Medical and drug research and FDA oversight did not save these women, because the medical community did not *see* them while it acted upon them. Their bodies were *out of place.*

Perhaps the problem has never been seeing a forest or a woman's body more clearly, through the sharpening lens of better and better science. Perhaps the problem has always been an arrogant un-will to see them differently. Now the world faces a century-long accumulation of such histories, entangled with each other in ways people also resisted seeing. But as Nancy has been chronicling in her recent books—*Sustaining Lake Superior: An Extraordinary Lake in a Changing World*, and this poignant volume, *Climate Ghosts: Migratory Species in the Anthropocene*—the tangles are now visible. They reinforce, and are reinforced by, the ferocious specter of our climate crisis.

Human-caused climate change is the great struggle of our time. We have the science to guide us toward a more humane (anthropocentric) *and* a more earth-friendly (biocentric) future. Yet we now know that science alone will not prevail, that its dominant paradigms cannot save the world. *Climate Ghosts* accepts this reality. Instead, the book asks us to begin seeing differently.

In *Climate Ghosts*, the beings whose fates we follow through a warming world are woodland caribou, lake sturgeon, and the common loon. Migratory species all, they are at once our ghosts, our guides, our collaborators, our most ancient new eyes. From thawing northern homelands, they take us into fields of fossil fuels, through clouds of synthetic toxins, among the brutalist structures of an industrial society risen out of Euro-American settler colonialism. These divine ghosts guide us elsewhere, too. Their complex historical relationships with Native peoples, far from being "past," vibrate with life in the present, as Indigenous communities initiate habitat restoration, introduce new approaches to fisheries and wildlife science, innovate legal protections for nature, and build pathways toward justice and amends.

When it comes to the natural world, western society has anointed western science as the aspirational pinnacle of knowledge. But the emerging concept of "two-eyed seeing" offers an alternative paradigm, and its influence is well underway. Two-eyed seeing flattens the pinnacle, so that traditional ecological knowledge has standing with academic science. Perhaps, *Climate Ghosts* encourages, if we learn to look through binoculars with both lenses, we might see woodland caribou galloping, lake sturgeon swimming, common loons flying toward hope in the Anthropocene. Our first hope, of course, would be simple survival for these sacred siblings and multitudes of other life (and this survival is not assured); but afterwards, maybe renewal, and then flourishing.

Mammal, fish, bird: these are the tragic climate stories, the wise cultural kinships, the hopeful alternative futures *in place*, that this book explores with open eyes and heart. Nancy Langston asks early on, "Can attending to history, in particular Indigenous history, help us devise better restoration strategies?" After each essay her question hovers and haunts. Can settler peoples learn from and work in humility alongside those who are better equipped to nurture and honor our relatives? Can capacious, tenacious love restore the song of the loon?

Lynne Heasley   NOVEMBER 2020

LYNNE HEASLEY is author of *A Thousand Pieces of Paradise: Landscape and Property in the Kickapoo Valley* (2012) and *Border Flows: A Century of the Canadian-American Water Relationship* (2016). Her current book is *The Accidental Reef and other Ecological Odysseys in the Great Lakes* (forthcoming 2021).

n the early 1990s, I spent two years studying Laysan and black-footed albatrosses caught as bycatch in the squid driftnet fisheries. The albatrosses had become entangled in miles of plastic threads, drowned as they tried to pluck out squid eggs to feed their chicks. Albatrosses are enormous birds with wingspans of up to 11 feet, and these graceful wings allow them to fly thousands of miles in a single flight. They breed on the leeward Hawaiian Islands and feed in the North Pacific Ocean, a place as remote from industrial civilization as any spot on earth.

My colleagues and I examined the stomachs of hundreds of albatrosses. Much to our surprise, nearly every bird we dissected contained plastic fragments in its stomach. Styrofoam was the most common contaminant. Many of these birds had incomplete molt patterns in their primaries—their main flight feathers—which reduced their ability to complete their long flights and breed each year. Parasites clung to their throats, indicating that their immune systems had faltered.

All those bits of plastic we were finding in the birds' stomachs may have been leaching toxins which led to immune and reproductive failure. A toxic sponge of trash was gumming up their bodies, even in the most remote of environments distant from human industry.

The abundance of plastics in human bodies, in water bodies, in wildlife bodies, highlights that synthetic chemicals found in the air, water, and soil are now been being detected *within* us. The chemical composition of our bodies and wildlife bodies is being altered in ways that reflect the transformations of our everyday environments. No matter how remote we think we are, no matter how wild the wildlife we encounter seem to be—we are all intimately entangled in the Anthropocene, an era where humans have become the dominant force on the planet, wreaking havoc with fossil-fuel extraction, toxic debris, and rapid climate change.

Climate Ghosts explores the histories and possible futures of three iconic migratory species of the Upper Great Lakes: one bird (the common loon); one mammal (the woodland caribou), and one fish (the lake sturgeon). I chose these three species because human cultures have formed close relationships, both material and spiritual, with each in the boreal north. Each has been the target of decades of restoration, yet their populations continue to fall. They are on the verge of becoming ghosts on the landscape, with dwindling numbers of their populations able to persist. Why?

In these essays, based on the Mandel Lectures in the Humanities that I gave at Brandeis University in 2019, I try to answer this simple question. I argue that the complex relationships different human cultures have developed with migratory wildlife are being rapidly transformed in the Anthropocene. Focusing on climate change and its interrelationships with toxics and habitat loss, I ask: How have the relationships between humans and other migratory animals been influenced by Euro-American settlement, energy extraction, industrial development, and climate change? How do animal migrations influence the mobilizations of toxic materials into distant spaces, and how does climate change in turn affect animal mobility? How can attending to history, in particular to Indigenous history, help us devise better restoration strategies?

Two chapters explore the challenges faced by woodland caribou in a warming world that favors their predators. Caribou evolution has been profoundly shaped by an ecology of fear, for they are vulnerable to wolf, bear, and lynx predation. Migration was one strategy they evolved to deal with the threat of predation. Grace, speed, wariness, an ability to thrive in the harshest winters and deepest snows—and a willingness to become partners with humans and become semi-domesticated—were strategies caribou across the circumpolar north evolved to persist in the presence of predators. Yet now their populations are dwindling across North America, in part because industrial extraction has altered the landscape to the point where they can no longer escape from predation. We celebrate the return of wolves, and we honor the devoted efforts of many wildlife biologists who reversed centuries of predator persecution. But the irony is that the species that wolves prey upon, particularly caribou, are now faltering. To save woodland caribou, wolf culls or caribou translocations—at least in the short term—may be necessary. If woodland caribou become ghosts over much of their range, people's memories of them may fade, and they might no longer be willing to do the hard, heartfelt work necessary for restoration.

The next migratory species we encounter is the lake sturgeon, a fish once on the verge of ghostliness, extirpated from a large fraction of its former range, yet still powerful in memories, stories, and art. This ghost story is more hopeful, for Indigenous communities have taken a key role in sturgeon restoration, and the results are promising.

The final species is the common loon, a long-distance migrant much beloved by northerners. Most people in the north have no idea that loons could become extinct, because we see our favorite pairs return to their lakes, year after year. More than a quarter-million common loons still breed in North America, so surely they're fine? But as long-distance migrants, like other waterbirds, they are vulnerable to environmental chaos in multiple places, not just at their breeding sites. Energy production on shores far distant from their northern breeding lakes affects them. They bear the ghostly traces of industrial extraction—mercury, PCBs, oil, lead—in their bodies, and those toxics are killing them.

Traditional perspectives within the humanities have placed humans at the center of the story and viewed humans as exceptional. In contrast, research within the environmental humanities focuses our gaze on the agency and interconnectivity of all things, displacing the human as the core source of value at the center of the universe. Environmental humanities doesn't give up on humans as a core part of the story, but instead sees humans as part of complex relationships—what theorist Bruno Latour has called "hybrid networks." The material and the cultural weave our world together, yet these weavings have often become invisible to us—they are, in effect, ghosts that haunt our attempts to heal relationships within a traumatized world.

My core argument in this book is that relational ways of knowing are critical, for we live in multispecies relationships. Humans are never all alone, making meaning in an empty world. Indigenous communities in North America have created collaborative relationships with loons, woodland caribou, and lake sturgeon, blending science and cultural practices in "two-eyed seeing." Such visions are critical for ethical restoration of ghost species in changing landscapes if we want to sustain hope in the Anthropocene.

ACKNOWLEDGMENTS

am grateful to Brandeis University and the Mandel Center of the Humanities for their generous invitation to offer the 2019 Mandel Lectures in the Humanities. My hosts at Brandeis, particularly Professors Ramie Targoff and Brian Donahue, were welcoming and helped to create an intellectually engaging and exciting visit. They offered me a priceless opportunity to try out new ideas with a receptive and lively audience. Staff at the center, particularly Diana Filar and Mangok Bol, did an excellent job of ensuring that my visit to Brandeis went smoothly. The faculty, students, and community members who attended the three formal lectures, seminar, class, and other events sparked lively discussions, and their perspectives have informed my essays.

I was fortunate to spend fall 2019 as the Mellon Visiting Scholar at the Center for Environmental Futures at the University of Oregon, where I worked on revising the lectures into essays. Professors Stephanie LeMenager and Marsha Weisiger have created a wonderful community of scholars interested in the environmental humanities. Fellows and postdoctoral scholars at the center, particularly Hayley

Brazier, Rebekah Sinclair, Allison Ford, and M Jackson were stimulating companions on field trips, hikes, and conversations at local pubs. Professor Mark Carey invited me to join his laboratory discussions for the fall, and these meetings offered valuable opportunities to engage with a fascinating community of students and faculty.

During the winter of 2020, I was awarded a Fulbright Research Chair in the Department of Geography and the Environment at Lakehead University in Thunder Bay, Ontario. Until coronavirus cut the semester short, the community of scholars and students at Lakehead University were most welcoming. In particular, Professor Brian McLaren in the Natural Resources Management Department was an excellent co-host. He invited me on field trips to various sites around the region, and he generously involved me in seminars and meetings with students, community members, and agency staff. I learned an enormous amount from our conversations, particularly on a long snowy drive to Wawa and back, and on our various snowshoe excursions. Bill Peterson, retired wildlife biologist in Grand Marais, Minnesota, was generous with conversations about caribou sightings along the North Shore.

In Wawa, Ontario, many people were kind enough to share their perspectives on woodland caribou management. Christian Schroeder invited me in 2018 to meet with Wawa residents concerned about caribou extirpation, and he provided numerous photographs and textual records. Gordon Eason, retired from the Ontario Ministry of Natural Resources, had initially met with me in 2010 to talk about caribou, moose, wolves, and forest management. In both 2018 and 2020, he spent a great deal of time with me sharing his experiences (and meals), leading a snowshoe excursion, and welcoming me to the community. He was also kind enough to review the caribou chapters of the manuscript with great care. Leo Lepiano of Wawa, formerly staff with the Michipicoten First Nation, was equally generous with his time and expertise. Aaron Bumstead, Director of Lands and Economic

Development with the Michipicoten First Nation, welcomed us to a long discussion about the challenges of woodland caribou restoration when we visited Wawa in February 2020. In Mercer, Wisconsin, wildlife biologist Jeff Wilson shared his expertise on loons with me.

A sabbatical leave from Michigan Technological University gave me the time necessary to revise the lectures into this book, with additional funding from the Fulbright Program and the Andrew Mellon Foundation. A National Science Foundation STS Research Grant (1921911), "The New Mobilities of the Anthropocene: Animal Migration, Infrastructure Development, and Wildlife Population Management," funded research and writing time on the caribou and loon chapters.

My editor at Brandeis University Press, Sue Ramin, has been a pleasure to work with, while Lillian Dunaj has offered able assistance with manuscript preparation. Cartographer Bill Nelson did an excellent job preparing maps.

Above all, I am grateful to my husband, Frank Goodman, for his love, encouragement, and willingness to hike, camp, and paddle in many of the sites mentioned in this book. I am lucky to share my life with his, and we're both endlessly fortunate to live on the shores of Lake Superior.

I thank *Agate Magazine* for permission to use some material in chapters 2 and 3 that first appeared in my January 2019 article, "Will Woodland Caribou Survive in the Lake Superior Basin?" Several paragraphs on Mongolian reindeer herders were initially published in Nancy Langston and Kate Christen, "Conservation Policies Threaten Indigenous Reindeer Herders in Mongolia," *The Conversation*, October 10, 2019. I thank *American Scientist* for permission to use material in chapter 3 that originally appeared in "Mining the Boreal North," *American Scientist* 101.2 (2013). K. Brosemer coauthored a forthcoming essay on loons with me, and I thank her for her insights.

CLIMATE GHOSTS

ONE

Ghosts
in the
Anthropocene

n 2007, chemists Will Steffen and Paul Crutzen and environmental historian John McNeill published an influential paper stating that we are now living not in the Holocene, as we all learned in geology class, but rather in the Anthropocene.[1] Humans have become the dominant force on the planet, overwhelming even the geologic processes of erosion and mountain building that defined earlier geological eras. As John McNeill and Peter Engelke declare in *The Great Acceleration*: "The period from 1945 to the present represents the most anomalous period in the history of humanity's relationship with the biosphere. Three-quarters of the carbon dioxide humans have contributed to the atmosphere has accumulated since World War II ended, and the number of people on Earth has nearly tripled. So far, humans have dramatically altered the planet's biogeochemical systems without consciously managing them. If we try to control these systems through geoengineering, we will inaugurate another stage of the Anthropocene. Where it might lead, no one can say for sure."[2]

While provocative, the concept of the Anthropocene can seem at odds with recent environmental humanities perspectives, which try to displace humans from the center of the story. The Anthropocene posits that we are possibly more important—or at least more hubristic—than ever before. Scholar Mick Smith warns us of the dangers of "uncritically accepting the term Anthropocene, for some may come to regard it as a badge of honour that (en)titles a new epoch of human technical mastery over the planet. There are, for example, those who refer to themselves as 'geo-engineers,' neo-Promethean fantasists who have learnt little or nothing from the failures of past attempts to provide technical fixes to ecologically and socially complex problems."[3] As humanist Claire Palmer notes, "Indeed, one potential problem with the idea of the Anthropocene is an exaggerated sense of the power of human agency. Much of what we think of as human impacts on Earth (e.g. climate change) is a combination of human actions and non-human processes, which humans can't control, and don't fully understand."[4]

Politics are core to understanding environmental change, but the concept of the Anthropocene may obscure the social, historical, and economic processes that have created rapid transformation. As Smith notes, "it is not humans per se that are responsible for the scale of this climatic and ecological impact but certain ways of organizing human societies that have become both divorced from ecological considerations and global in extent."[5] While I recognize the validity of these critiques, the concept of the Anthropocene still does useful work for us, because it helps focus our gaze on the enormity of the disruptions challenging life on earth.

As an environmental historian, I use a historical lens to examine the interconnected, hybrid relations between nature and culture over time. Geographer Robert Wilson argues that a hybrid approach to animal migrations involves reconsidering animals, migration corridors, and human infrastructure such as roads, dams, and pipelines not as separate categories, but as hybrid networks. Hybrid systems collapse the traditional boundaries between human and nonhuman, cultural and natural. Historian Sara Pritchard asks: What happens when we really consider "the deep entanglement of people and the environment"?[6] *Climate Ghosts* follows Wilson and Pritchard's challenges, focusing not just on human policy decisions, but also on the hybrid ecological, toxicological, and cultural processes that weave together natural and human histories in the Anthropocene.[7]

Why Consider the Agency of Other Creatures?

One of the core arguments in this book is that other animals have agency—but what does that mean, and why does it matter? Agency has traditionally been conceived of as self-conscious, intentional autonomy, the ability to determine one's own future, to plan an intended outcome and act to bring about one's intent. By this definition, most other animals lack agency because they have little intentionality about the future. As theorist Amanda Rees writes, "one aspect of humanity has seemed indisputably ours alone: our capacity for self-conscious agency; that is, our ability not just to act on and influence the world, but to do so deliberately and reflectively."[8] But this concept of agency is problematic, as Rees argues, for it privileges a certain narrow type of power, while marginalizing the contributions of women, people of color, and other species. It assumes that human intentions actually do determine the outcomes of action, and it ignores the myriad of other processes that shape change—both environmental and human. As Rees points out, our sense of intentionality and self-direction is often an illusion. Language, art, the transformation of forests into farms, the domestication of other species, our very DNA—what if these really weren't grand human projects forced upon an unwitting natural world, but rather multispecies collaborations and negotiations? What we think of as purely human achievement "might actually have emerged from our relationships, past and present, with different animal species."[9]

Western cultures have seen nonhuman animals as profoundly "other"—fundamentally peripheral to the grand narrative of human accomplishment, except as useful tools. In this view, only humans have rational intelligence and can use that to transcend nature.

While these Enlightenment beliefs have shaped modern conceptions of humans as rational autonomous agents, the dualism that underlies these beliefs has much deeper roots. Theological traditions in western culture have largely assumed that humans are fundamentally different than other animals, for humans alone have the potential to reason and to transcend nature. When asked the question "Does the universe have a purpose?" David Gelernter answered that the purpose of the universe was for humans—and humans alone—"to defeat and rise above our animal natures; to create goodness, beauty, and holiness where only physics and animal life once existed; to create what might be (if we succeed) the only tiny pinprick of goodness in the universe—which is otherwise (so far as we know) morally null and void. . . . When we seek goodness and sanctity, we defy nature."[10]

Enlightenment scientific traditions, from Descartes on, built on these dualistic religious frameworks of human exceptionalism, adding to them the belief that other animals are mere machines, lacking meaning. As mechanical constructions, Descartes and his followers argued, nonhuman animals could not self-consciously reason, and so they lacked feeling, self-awareness, and agency.[11] Euro-American colonialism in North America rested on these same core assumptions: the boundary between human and nonhuman is absolute, and other animals are tools to be used by humans, not kin.

From Indigenous perspectives, these beliefs are bizarre. In the chapters that follow, I explore some of the ways that different Indigenous cultures in North America have understood the rest of the world to be radiant with meaning—to be in relation with us and each other. For these cultures, restoration must be centered not just in increasing population numbers, but also in restoring relationships across species and across watersheds. One philosopher asked me: "How can you ask us to rethink human-animal relations? That's at the core of western philosophy; it's not going to happen." This book is a response to that challenge. I argue that Euro-American settlers in North America have much to learn from Indigenous cultures. Restorationists don't need to invent a new philosophy; they need to learn from Indigenous communities who have already articulated relational ways of knowing.[12]

In the last several decades, research by animal behaviorists has found abundant evidence which shows that the dichotomies between animal and human are culturally constructed. New research by biologists shows that many nonhuman species, particularly primates, whales, and elephants, can create rapid cultural innovation. More than two decades ago, in 1999, an international survey of wild chimpanzees published in *Nature* described thirty-nine distinct behavior patterns—showing that separate communities of chimpanzees, even in the same environment, innovate different social customs. Apes learn to medicate themselves with herbs. Chimpanzee mothers show their young how to use stones to crack nuts. Monkeys teach their siblings how to wash sweet potatoes in the ocean. Even fear needs to be learned. When young monkeys raised in captivity were shown live snakes for the first time, they remained utterly unafraid until they observed their elders reacting with fear. Crows and ravens keep track of cheaters and punish those individuals who violate their social norms. Whales teach their children songs and sing throughout the ocean, filling the watery world with meaning that only they can grasp. These are but a few of the many examples of a nonhuman world imbued with cultural meaning.[13]

What we see when we look at other animals reflects our own cultural perspectives. Euro-American cultures have long framed other animals as distant from human sources of meaning—philosophical, cultural, religious, intellectual. But with growing concerns about climate change, anthropologist Julie Cruikshank argues, westerners are returning to a sense that humans are engaged in intricate relationships with other animals as well as nonliving beings, such as rocks and climate. She locates similar patterns in traditional Indigenous narratives. Cruikshank writes: "The balance of evidence suggests that our human ability to come to terms with global environmental problems will depend as much on human values as on scientific expertise, especially in an increasingly alienating and uncertain world. Science and local knowledge have come to be seen as polar opposites, yet mutual stereotypes share similarities. . . . In oral narratives from this region, we hear stories about the importance of human agency, human choice, human responsibility, and the consequences of human behavior, and it is here that one of their contributions to climate change research may lie. Narratives underscore the social content of the world and the importance of taking personal and collective responsibility for changes in that world. . . . In the past, then, things and people were always entangled. In the future, they will be more entangled than ever before."[14]

Conservation Ghosts

I focus this book on three ghost species in the Great Lakes watershed—woodland caribou, common loons, and lake sturgeon. Ghost species are those that have not gone completely extinct, although they may be extirpated from a particular area. Their traces are still present, whether in DNA, in small fragmented populations, or in lone individuals roaming a desolate landscape in search of a mate. We can still restore them, if we make the hard choices necessary for them to survive. We catch glimpses of these ghosts in memories, in dreams, in petroglyphs on rock faces, as we paddle through lakes that are still extraordinarily lovely—but strangely depauperate, barren of the blooming, buzzing diversity that once filled each lake with song. We see these ghosts in the names of coffee shops. Take Caribou Coffee, one

MAP 1.1 The Great Lakes watershed in North America. *Bill Nelson*

of the region's popular suburban café chains. Yet most of the people who grab a latte from the friendly staff at Caribou Coffee don't know that caribou once walked where suburbs now sprawl. What can these ghosts teach us?

I've been thinking about ghost species since the 1990s, when I was in graduate school in ecology and living in Seattle. Those years, I spent as much time as I could backpacking in the North Cascades. One night, reading by headlamp in my tiny tent at the edge of a cliff in the Cascades, my elderly dog snoring by my side, I read these words in Edward Grumbine's popular book about conservation biology, *Ghost Bears*: "The grizzly embodies all the problems and promise of the biodiversity crisis in the North Cascades. The bear has been protected by the Endangered Species Act since 1975, yet until recently no study had been initiated to determine the status of the species in the region. There are certainly grizzlies in the North Cascades."[15]

"Really?" I thought, sitting up. I had always hung my food well away from my backpacking campsite, cautious about black bear encounters. But grizzlies? Much as I was brave about backpacking without a human companion (just so my dog was at my side), grizzlies made me pause. Backpacking in grizzly country—with the fear that goes with it—is a very different activity than backpacking in a landscape without them.

Grumbine describes how, as late as 1800, perhaps 100,000 grizzlies still inhabited North America, from Ontario in the east, to the Pacific Ocean in the west, and south into Mexico. In what's now the Lower 48 states, 50,000 grizzlies roamed.[16]

But Euro-American incursions into the North Cascades, as across North America, led to a crash in grizzlies. He describes four waves of decimation: trappers in the early to mid 1800s; miners who came soon after; ranchers from the 1850s onward; and intensive logging beginning in the 1870s. The result? When Grumbine wrote *Ghost Bears*, 99 percent of grizzlies south of Canada had been destroyed, and less than a thousand remained in the Lower 48.[17] Only ghost traces of grizzlies persisted in the North Cascades.

Of the web of connections between humans and bears, Grumbine writes: "The bear has, so far, slipped through the net of scientific inquiry we have thrown over the mountains. . . . The grizzly bear is a ghost in the Great North Cascades, off trail, beyond the campfire's light, living in the hills unseen, playing at the edge of dreams to some, nightmares to others. In the deep past, humans developed a complex relationship with the bear based on respect."[18]

Grumbine calls for renewed respect for grizzlies, for attention to

Indigenous knowledge and, above all, for a biodiversity perspective that focuses not just on individual species-by-species restoration, but on ecosystem management. Ghosts species, he argues, cannot be ignored: we should do all we can to stem their decline, before they lose their special place in human imaginations, and communities therefore lose the will to restore them.

Grumbine's argument builds on a developing consensus in the conservation biology literature. The restoration of entire watersheds, of ecosystems in all their complexity, needs to be part of any comprehensive strategy for endangered species, Grumbine insists.

In the decades that have passed since *Ghost Bears* was published, conservation agencies around North America—federal, state, and provincial—have begun to try to address broader ecosystem processes, not just individual species counts. Ecosystem management has become an accepted part of the philosophies of the U.S. Forest Service, Park Service, and Fish and Wildlife Service, state departments of natural resources, and provincial ministries in Canada. These agencies now invite extensive public input, calling for participatory management, long-term vision, and attention to historic ranges of variability—not just restoration to one static moment in an imaginary past.[19]

But how well has this worked? The efforts to restore woodland caribou, common loons, and lake sturgeon that I describe in the chapters that follow suggest the results are mixed. Ecosystem management and collaborative participatory management are important perspectives—but they can also provide ready excuses to delay hard decisions that might anger key constituents.[20]

The effects of predators in an ecosystem are more profound than we can easily imagine, even when we live and work and play in the landscape where they roam. Predators create what biologists now call "a landscape of fear."[21] Having to be alert to the presence of predators shapes an herbivore's day-to-day actions, and it shapes their evolution. The grace of an antelope, the wariness of a woodland caribou, the speed of an elk—all these didn't just happen on their own. They evolved because prey animals were in an evolutionary dance with their predators.

When we remove predators from the landscape, prey species become ghosts of their former selves over evolutionary time. But if we have too many predators in a landscape so altered by human activity that their prey cannot escape, then those prey become literal ghosts: impossible to sustain. Part of the challenge of modern conservation is retaining predators such as grizzly and wolves on the landscape, without destroying the ability of their prey species to persist. One of the core challenges for woodland caribou now is surviving in a world where their predators—wolves, bear, and lynx—have been partially restored, even as those habitat connections that allowed caribou to coevolve with predators have not been repaired.

In the summer of 2019, my husband and I camped on a road trip to Oregon, where I would spend a term as the Mellon Visiting Scholar in Environmental Humanities at the University of Oregon. Another elderly dog snored by our side, this time our fourteen-year-old pit bull, Vanya. When we camped for just a few nights in grizzly country in the Selkirk Mountains, in a landscape where reproducing populations of woodland caribou had recently been extirpated, I remembered that particular attention required when camping with predators who might decide you are dinner—particularly when one of their ancient prey-companions, the woodland caribou, has been lost. Only three lonely female woodland caribou still walked the southern Selkirk Mountains, a population unable to reproduce. Just before a species goes extinct, those last few individuals of the species (called *endlings*) are ghosts of a kind, living alone in a world where their relatives have vanished and their possible mates are dead.[22] These last caribou were endlings in the Selkirks, unable to sustain any kind of healthy predator-prey relationship with the grizzly bears and wolves who were returning to the mountains (see plate 1).

Ghost stories about extinction ultimately ask us to consider these questions: What can we learn from the unresolved traumas of historic colonization, and the eradication of Indigenous communities, wildlife, and ecosystems, to keep our fellow creatures and ourselves from extinction in a warming, politically fractured world?

Woodland
Caribou Histories
in the
Upper Great Lakes

n January 2018, the population of Lake Superior woodland caribou
nearly went extinct (see plate 2). Once the dominant deer species
across the north, woodland caribou (*Rangifer tarandus caribou*)
had roamed from Hudson Bay to Mackinac Island in Michigan.
They had persisted for more than 1.6 million years in North America,
thriving through the dramatic environmental fluctuations of the Ice
Age. When glaciers encroached, they found refuge in the southern
mountains of Appalachia; when temperatures warmed, they moved
northward with the melting ice to fill the forests and islands along
Lake Superior.[1]

But in 2018, fifteen of the last individuals were trapped on an island on the Canadian side of Lake Superior with a burgeoning wolf
population. Local community members in Wawa, Ontario, led by Leo
Lepiano of the Michipicoten First Nation, retired ministry biologist
Gordon Eason, and cottager Christian Schroeder, urged the Ontario
Ministry of Natural Resources and Forestry to save these remaining
woodland caribou.

After media attention to the caribou's plight, the ministry acted. In three dramatic interventions in January and February 2018, biologists captured caribou in large nets and helicoptered them to wolf-free islands. Nine were taken back to the Slate Islands archipelago, where wolves had vanished after eating all but two of the caribou, and six others to Caribou Island, 37 miles (60 kilometers) south of the Canadian mainland.

As Senior Policy Advisor Katherine Olejarz explained in an e-mail, this took some time because the team had to "ensure the correct permits, safety procedures and logistical plans, complete with contingency plans" were in place, and then they needed "to execute a complex translocation in a remote island setting located on Lake Superior. In this type of translocation, weather is a big factor with human and animal health being the number one consideration. Suitable ice and snow conditions were required to safely land aircraft and capture the caribou." According to Olejarz, by early 2019 the translocated caribou "appear to have settled in."

The caribou are reproducing, giving hope for a possible population recovery. Along a remote beach on the Slate Islands, a biologist found the prints of a caribou calf in the summer of 2018. Photos from camera traps showed at least one calf as well.[2]

For now at least, extinction has been averted. But what lies in wait for woodland caribou in a warming world? Can last-minute interventions continue to save them from the brink? Does it make sense to save a few last woodland caribou, if they're doomed to extinction in the Anthropocene anyway? What do ghosts tell us about our future?

Woodland Caribou Ecology

Woodland caribou are part of a globally distributed species, *Rangifer tarandus*, which includes reindeer in Eurasia, barren-ground caribou across the North American Arctic, and woodland caribou in the boreal subarctic. Members of the Cervidae (deer) family—which includes deer, elk, and moose—caribou thrive in a variety of habitats. They are a migratory species that can cover vast distances across the tundra or adapt to much shorter migrations in forests. Barren-ground caribou and Eurasian reindeer are famous for migrations covering more than a thousand kilometers, while Lake Superior woodland caribou have shorter movements from wintering to calving ranges.

The woodland caribou's range once extended across the entire northern forest, from the boreal forests that stretch from Hudson Bay to the Great Lakes, south to the mixed hardwood forests of the lower peninsula of Michigan, east to Newfoundland, north to the boreal Arctic. But in the United States, woodland caribou now persist only in place names and memories. Across Canada, they've retreated from roughly half their nineteenth-century range.

In May 2018, woodland caribou were declared functionally extinct in the United States. The last remnant population in the Selkirk Mountains of Idaho had dwindled to three lone females. When my husband, dog, and I camped at the edge of the Selkirks in September of 2019, we slept in a land threaded with caribou ghosts. Petroglyphs, stories, place names, all speak to their historic presence on the landscape, and to the myriad ways that human and caribou lives were woven together, shaping both species.

Within the Great Lakes basin, they survive only in a few tiny populations along the Canadian north shore of Lake Superior. By 2017, across Canada, more than half of the fifty-seven distinct populations of woodland caribou across Canada were in steady decline. As Hillary Rosner writes, where oil and gas development overlaps with caribou ranges in Alberta, those populations are "shrinking by half every eight years. Scientists now predict that nearly a third of Canada's boreal caribou could disappear within the next 15 years."[3]

One of the few large megafaunal species to expand rather than go extinct in the Late Pleistocene, caribou survived repeated glaciations by moving to ice-free refugia. Boreal woodland caribou retreated to

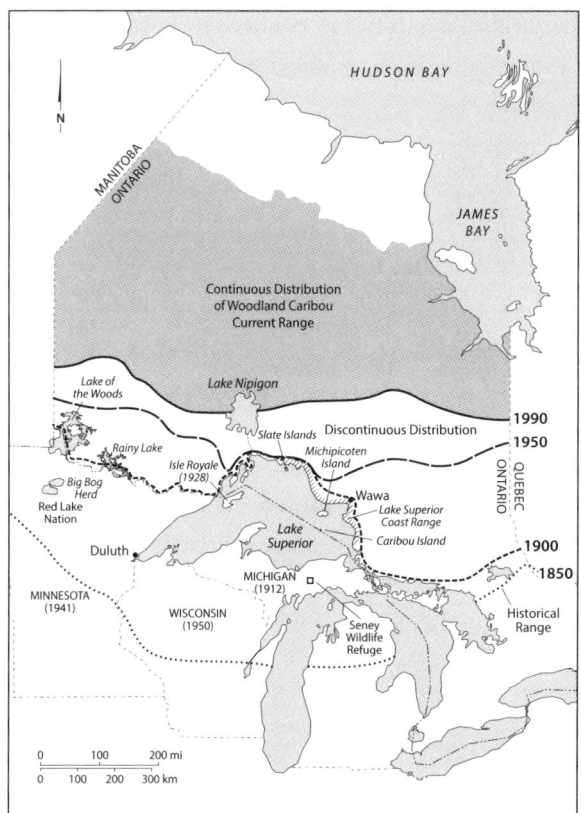

MAP 2.1 Woodland caribou have steadily retreated
from the upper Great Lakes. *Bill Nelson*

the Appalachian Mountains, while Eurasian reindeer moved to what's
now Italy, France, and Spain. At the end of the last glacial maximum,
as the glaciers retreated, caribou chased the melting ice north, exploring the new environments that opened up as the climate warmed
and expanding their range across the circumpolar north. Populations
had increased by the end of the Pleistocene, even as most other large
mammals (with the exception of brown bear and tundra muskox)
went extinct. Migration was central to the caribou's post-Pleistocene

resiliency. Each time the ice retreated in an interglacial period, caribou followed the melting ice north, expanding into new habitats across a diverse, warming landscape.

But now, after millennia of adaptation to changing environments and changing relations with people, woodland caribou face grave challenges across North America, not just in the Lake Superior basin. Is climate change dooming woodland caribou? Or are managers using climate change as an excuse to avoid making difficult policy decisions that could save the caribou but antagonize industry and environmental groups?

In the Lake Superior basin, a genetically distinct population of woodland caribou developed, ranging as far south as the Lower Peninsula of Michigan and as far north to Hudson Bay in Canada. Woodland caribou were widespread across the northern forest, but they were never particularly abundant in any place according to caribou researcher Thomas Bergerud, for low caribou numbers helped keep their wolf predator populations similarly low. Unlike the much larger moose, which may stand their ground and kick a wolf trying to eat a calf, caribou flee from predators. They find refuges from predation in deep snow, windswept barrens, dense bogs, or rocky coastlines, where wolves falter.

In Ontario during the mid-twentieth century, biologists with the Department of Lands and Forests pieced together their understanding of woodland caribou movements and wolf predation by interviewing Indigenous hunters. In 1959, for example, D. W. Simkin reported that "Michel Hunter, the chief at Winisk, and a very good hunter, said that the coastal caribou move out to the coast in April or March and move back to the timber in December depending on early or late freeze up." He added that all his Indigenous informants reported that wolf predation was generally quite low, but when wolf packs came upon caribou, they often made numerous kills before moving on.[4]

Predator avoidance was a strategy that served woodland caribou

FIGURE 2.1 Woodland caribou were once widespread across the northern forest before industrial development. (Photo from William Berryman Scott, *A History of Land Mammals in the Western Hemisphere*, 1913.)

well in North America for over a million years, but now they are vulnerable to predation if their migration routes are cut off by development, or if predator populations increase with human disturbance. And many forms of human activity do increase predation. Railroads, logging roads, forest conversions, and wetland drainage have offered easy access for human and canid predators. Moose and white-tailed deer have expanded their range into caribou territory, for the edge habitat left by forestry serves them well. Deer and moose in turn invite higher wolf populations, while deer also spread a parasitic brainworm (*Parelaphostrongylus tenuis*) that kills caribou.

Caribou are creatures of the boreal north. Ecologically, boreal ecosystems share certain constraints. They develop in places with extremely cold winter climates and short growing seasons, on landscapes only recently freed from glaciation. These are disturbance-prone ecosystems marked by change, not by stasis. Cold as they are, they are still extraordinarily rich in water, forage, minerals, and animal life. When the snow geese come to the Arctic in their millions, when the caribou herds migrate to their breeding grounds and, above all, when the mosquitoes come, the word *abundance* takes on new meaning.

Animals that live in the boreal forest have evolved different strategies for surviving the harsh climate. Some hibernate, some grow a warmer winter coat and stay put, while still others, such as caribou, migrate into winter habitats—not places that are warm and Floridian, but cold places that provide better access to food, protection from harsh winds, or escape from predation.

Caribou are well-adapted to frigid winters. Their fur is thick and their hairs are hollow, providing excellent insulation. Like many waterfowl, they have heat exchangers in their limbs so the blood in the veins, which has cooled significantly as it comes up from the hooves, is heated up by the arterial blood before it reaches the heart again. Reindeer nostrils are dense with blood vessels, so the air they breathe is warmed up and heat loss via exhaled air is minimized. Even with these elaborate adaptations, many caribou may starve in particularly severe winters, while many cows will miscarry. Caribou migration patterns reduce some, but not all, of their vulnerability to climate fluctuations.

In other species of the deer family, such as moose or mule deer, the males grow antlers and shed them each year, while the females lack antlers entirely. Caribou are different. Both bulls and cows grow antlers, and females retain them during the winter, allowing them to drive off the bulls from winter pastures, when the cows are pregnant and need all the energy they can get. The cows do not lose their antlers until they have given birth to calves in the spring. Each cow gives birth to a single calf, rather than the frequent twinning that occurs in moose and whitetail deer. During pregnancy and calving, they are particularly vulnerable to disturbances which can increase the risk of miscarriage.

During the summer, when there is little snow on the ground but plenty of mosquitos, biting flies, and midges, caribou move to places where they can find some relief from insects. On the buggiest days, they may huddle at the edge of melting snowbanks; other times, they might head up rocky slopes or to the outermost fragments of small islands. As the fall comes, they typically migrate to autumn pastures, where they eat vast amounts of grass and lichen, putting on the fat reserves critical for making it through a harsh winter. When the snows deepen in late autumn, they again migrate, this time into forests (where they can dig through the snow to reach their main winter food source: ground lichens of the *Cladonia* genus and various tree lichens. Such areas have also been the target of industrial forestry, which greatly reduces forest lichens and changes the patterns of snow buildup in the regenerating forests, making times tougher for the caribou.[5]

Woodland caribou have evolved many adaptations to the extreme cold and deep snow of boreal winters. They have two layers of fur: a fine underhair that helps trap heat combined with outer guard hairs that are hollow, acting like insulation. Their feet have four toes: two small dewclaws and two larger concave toes that offer support on boggy ground and crusted snow. Long hair covers the skin between the toes, keeping their feet warmer. The pads within their hooves change when winter comes, allowing them to resist the cold and move easily over snow. Their long legs can navigate deep snows without floundering. They have countercurrent heat-exchange systems that enable them to maintain much higher temperatures in their body core than in their legs. They have an extraordinary sense of smell, so they can sniff and then dig out ground lichens buried under 5 feet of snow. With their four-chambered stomach and complex digestive system, they can digest nutrients from lichen and rough grasses.[6]

During calving, they are vulnerable to predators, particularly wolves, bear, and lynx. But they have coevolved with these predators and, like other herbivores, they have developed numerous strategies that have allowed them to persist for millennia. Different members of the deer family have different anti-predator strategies. Moose stand their ground when a predator approaches and can often defend their calves. White-tailed deer, like caribou, are too small to defend against wolves; instead they hide their young, often have twins, and typically begin to breed at a young age. Many of their young die, but a doe can readily breed again over her lifetime. Woodland caribou have another approach: they have just one calf a year and invest heavily in that single offspring. To protect against predators, they disperse widely, keeping their densities low so it's less likely that a predator will find them.

At calving time, woodland caribou often retreat to rocky islands. They are far better swimmers than wolves or bears, and they move more easily on ice, so they can find refuge from predation on the smaller islands that wolves and bears typically have trouble reaching. They also can move more easily than their predators through bogs, lakes, and deep snow, so wetlands in particular can offer some protection.

While woodland caribou have been widespread in the northern forest, they were rarely abundant in any given place. This means that when caribou are the only deer species, wolf numbers stay fairly low, because there is simply not a large enough prey base to support expanding wolf packs. In an intact forest, woodland caribou can outrun wolves through deep snow. But when industrial forestry, extractive energy development, gas pipelines, snowmobiles, or even snowshoers enter the winter forest, the paths they create ease wolf passage. Wolves can now enter the refuge of the caribou, which means that caribou must then retreat ever further into the heart of the forest, until they simply run out of safe havens.

When other deer species, such as moose or white-tailed deer, move into caribou forests, that too spells trouble for caribou. When moose numbers grow large, wolf numbers increase as well, and those wolves may turn to caribou as an easier meal. When white-tailed deer invade the northern forest, they not only attract wolves but also infect caribou with parasitic brainworms. Deer are fairly resistant to brainworm damage, but those same parasites wreak havoc on moose and caribou.

Great Lakes Woodland Caribou

How do we know where woodland caribou and their predators were once found in the Great Lakes region? Multiple lines of evidence help create a fuller picture.[7] Archeologists looks for ancient remnants of bones, teeth, and antlers, often focusing on Paleoindian sites to reconstruct a picture of human-caribou relations in deeper history.[8] For example, a bone from a caribou was excavated in southeastern Michigan, recovered from a pit estimated to date back 11,000 years. This pit contained charcoal and chert fragments, evidence of human occupation.[9] Caribou hunting sites from nearly ten thousand years ago have been found on the bottom of Lake Huron, preserved when ancient lakes flooded the region.[10]

While archeological evidence provides fragments of the past, it's critical to interweave these perspectives with Indigenous narratives and knowledge, as Julie Cruikshank has done in her collaborations with First Nations communities in Yukon.[11] Stories from elders, such as those of Francis Pegahmagabow, an Anishinaabeg elder and member of the Caribou Clan, show how central caribou were for Anishinaabeg people along the Great Lakes well into the twentieth century. For example, Pegahmagabow met another member of the Caribou Clan from Thunder Bay, who recognized him as a fellow clan member and so gave him a medicine bag to protect Pegahmagabow from death in World War I.[12]

Scholars also turn to the written records of Europeans, who entered the upper Great Lakes region in the seventeenth century. European traders marveled at the diversity of deer species, including moose, caribou, white-tailed deer, and elk.[13] In 1658, for example, the fur trader Pierre Radisson mentioned seeing caribou as he travelled along the south shore of Lake Superior.[14] From these records, it's clear that woodland caribou in the Great Lakes region once wandered "as far south as the vicinity of the Twin Cities"; in Michigan, they made their way south of the Straits of Mackinac.[15]

Written records suggest that woodland caribou were widespread in the Lake Superior basin, finding their way even to remote islands. Fur trader Alexander Henry, partner in the North West Company, described Anishinaabeg life within the upper Great Lakes country in his memoir, *Travels and Adventures in Canada and the Indian Territories in the Years 1760–1776*. Henry grew impatient at the slow pace of fur trading, and he decided to try for quicker riches, venturing for gold rather than beaver.[16] In 1769, Henry sailed to what's now known as Caribou Island, 37 miles from the mainland, having heard tales from Native Americans about beaches of yellow dust (which he presumed to be gold), guarded by enormous snakes. This first attempt to cross the lake failed, so two years later Henry partnered with a man named Baxter. They formed a mining company, built a bigger barge, and sailed off again for the Island of Yellow Sand, hoping for a paradise of gold dust. After three days' sail, Henry and his crew landed on the island and leapt off the boat ready "to bravely fight off the giant snakes the Indians claimed inhabited its shores."

Much to his disappointment, he found no gold (or giant snakes). Instead he saw, in the forest near the water's edge, the "tracks of cariboux. Soon after I entered the woods three of these animals discovered themselves and, turning round, gazed at me with much apparent surprise. I fired at one of them and killed it; and at a mile farther I killed a second . . . . The day following I killed three."[17] By the end of a three-day stay, he bragged, the party killed ten more caribou, for a total of thirteen.

Henry hoped to find beaver but reported that "the only four-footed animal was the caribou and this, it is probable was first conveyed to the island on some mass of drifting ice. It was, however, no new inhabitant; for in numerous instances I found the bones of cariboux, apparently in entire skeletons, with only the tops of their horns projecting from the surface, while moss of vegetable earth concealed the rest. Skeletons were so frequent as to suggest a belief that want of food in this confined situation had been the destruction of many; nor is anything more probable; and yet the absence of beasts of prey might be the real cause."[18] In other words, there were no signs of bear, lynx, or wolves on the island. Without predators, Henry speculated, the caribou must have become so numerous that they decimated their food and their populations crashed—an interesting ecological interpretation.

Half a century later, in 1789, a nine-year-old Ohio child named John Tanner was captured, then adopted, by an Anishinaabeg family. Tanner spent years in what's now Minnesota with his adopted community. In his journals, published in the 1830s, he described an abundance of caribou throughout the region. He was invited to travel to a "large island in Lake Superior [Isle Royale], where, [a man] said, were plenty of Caribou and Sturgeon." There they found thriving herds of caribou.[19] Tanner wrote that, in his younger days, caribou were abundant enough that his family would head out on winter journeys with little packed food, assuming that they could always find a caribou when needed.

But when the caribou failed to appear, hunger threatened. Elders such as Tanner's adoptive mother, Net-no-kwa, would intervene with the spirit world, calling the caribou spirits forth to ensure a respite from starvation. Tanner observed that Net-no-kwa "tracks game in her dreams," allowing the family to survive harsh times.[20] These scenes suggest the close spiritual relations that the Anishinaabeg formed with caribou, as well as the importance of women elders in the community. Tanner's narrative is not just a portrait of traditional relations between caribou and people. It also illustrates the devastation wrought by the fur trade. It's a portrait of relations in rapid transition, as Louise Erdrich writes, "disrupted by the demands of large fur companies for huge

numbers of fur-bearing animals. Starvation haunted Ojibwa winters."
Tanner writes over and over again of the "all-absorbing task" to stay
alive during those times of famine.[21]

Less than twenty years after their arrival, Europeans were noting
declines in caribou throughout the Lake Superior region.[22] Devastation fell upon the Native Americans, as the wildlife they depended
upon fled. Smallpox and other diseases ravaged their communities
while violence from the Europeans caused terrible deaths. According
to W. W. Warren, who prepared a "historic sketch of the Ojibwa" for
Territorial Governor Alexander Ramsey in 1850, the Ojibwe who
inhabited the region from Minnesota's North Shore westward to
Lake of the Woods were by that time "miserably poor" and numbered
about eight hundred. Their winter food was "rabbits and reindeer." But
caribou could no longer be depended upon, and rabbits were equally
uncertain. Warren noted that "In the winter of 1849, there was a rabbit
die-off and 17 of these Indians starved to death."[23]

Caribou, wolves, bear, lynx and Indigenous peoples had managed
to coexist for millennia after the ice retreated, with predators and prey
ebbing and flowing across a complex landscape. But after colonists
from Europe settled, a familiar history unfolded. The first stressor
was intense hunting. Caribou were quickly targeted by elite white
recreational hunters, for caribou were perceived as sporting to kill
because they so cunningly evaded predators. Edward Samuels, one of
many hunting writers who extolled Canadian and American hunting
opportunities, wrote in 1897: "By many sportsmen of experience the
woodland caribou is given a higher place in the category of game
animals than the moose. His solemn, almost patriarchal aspect,—his
silent, furtive, whimsical ways,—his mysterious migrations from one
section of the country to another, which seem to be the product of
sheer restlessness rather than of reason or necessity,—his wonderful
speed and endurance in traversing the deepest snows of winter and
his capacity to thrive upon such evanescent and ethereal fare as the

reindeer-lichen, are among the factors which make the caribou an object of interest to all who have formed his acquaintance on his native heath. Like the moose he is a true child of the wilderness and intolerant of the presence of man."[24]

As Europeans and settlers expanded the fur trade, mining, and logging into the north woods, habitat for woodland caribou and refuge from their predators—particularly human predators—diminished. Many woodland caribou fled to islands or dense bogs and became known as creatures of wetlands. Some fled to high mountains. All tried to get away from people.

In 1869, for example, Campbell Hardy observed that formerly abundant caribou were giving way to moose, which were more tolerant of disturbance. "In many districts," he wrote, "especially those in which the existing southern limits of the caribou are marked, this animal is gradually disappearing, whilst the moose is taking its place."[25] Hardy noted that as caribou retreated, the help of a Native guide was required to hunt them: "the wonderful tact of the Indian is indispensable in a forest country, where the game cannot be sighted from a distance."[26] Hardy recognized that hunting alone was not the cause of caribou declines. Euro-American settlement, forest disturbance, and resource extraction were all deadly to caribou. "To a great extent this [loss of caribou] is the result of an increasing settlement of the country by man," he continued, because "the caribou is a great wanderer, and requires long and unbroken ranges of wild country in which he can uninterruptedly indulge his vagrant habits. Being moreover more jealous of the advance of civilisation than the moose, he is surely disappearing as his old lines of periodic migration are encroached upon and broken by new settlements and their connecting roads."[27] Hardy mourned that for "the wilder races" of Indigenous peoples in Canada, the caribou's "gradual disappearance must bring starvation and a corresponding progress towards extinction."[28] In this discourse of inevitable decline, both caribou and Indigenous peoples were framed as "wild" animals on the way toward extinction—sad, perhaps, but not something that whites could stop.

In 1877 biologist John Caton noted that caribou had been in steep decline since 1820. In fifty years, caribou had been "greatly diminished . . . in nearly all the countries where they were formerly quite abundant."[29] And he noted that they had retreated to wetlands, swamps, deep forests, and rocky shorelines—both to evade natural predators and human predators: "It frequents marshy and swampy grounds more than any other of the deer family . . . where it is well protected from pursuit. In the winter it resorts to the dense forests on higher ground."[30]

When caribou populations began to decline, many whites blamed the Native guides they depended upon for their own hunting adventures. Elite conservationists forbade Indigenous people to hunt caribou, even though some whites recognized that such action could lead to devastation for the Indigenous peoples who depended on caribou. In 1896, an elite hunting writer, Caspar Whitney, wrote with horror of what he saw as "the annual slaughter visited upon them" by the tribes, adding "If, therefore, the Indians starve because of unskilled hunting, it is only just retribution for their improvidence and rapacity."[31]

Many whites saw what they interpreted as Indigenous "rapacity" as part of some essential barbarism. One white hunter, Samuel Davis, complained in his popular book *Caribou Shooting in Newfoundland* (1894) about his guide "Indian Jim," whose "Indian instincts got the better of him" when the guide suggested that his client shoot a doe for meat, rather than focusing only on trophy bulls.[32]

Long after whites condemned the "improvidence and rapacity" of Indigenous hunting, those same white hunters celebrated their remaining opportunities to slaughter woodland caribou. For example, the province of New Brunswick allowed outsiders to kill game long after other provinces began to limit hunting. Edward Samuels wrote in 1897: "It is reasonably certain that no section of North America within convenient reach of the big-game sportsman, with the possible exception of Newfoundland (where the caribou attains his greatest perfection but where the laws are very stringent as to visitors), now offers facilities equal to those of New Brunswick for caribou hunting." Samuels noted that intense recreational hunting had devastated caribou and moose populations on the U.S. side of the border: "Maine is still pre-eminent in its supply of deer, but so great has been the invasion of sportsmen upon its hunting-grounds of late, that moose and caribou are now comparatively scarce, except in the Aroostook region." And because of that, Samuels speculated, the "persistent hunting, in season and out of season, which these animals have experienced of recent years in Maine has resulted in a considerable exodus to New Brunswick."[33] But rather than extrapolate from the experience of Maine's collapsing wildlife populations, Samuels urged elite hunters to hunt woodland caribou in Canada as long as they persisted. He wrote: "The game laws of New Brunswick are certainly liberal in all their features. They permit an open season for moose, caribou, deer, duck, woodcock and snipe."[34]

Such disregard for hunting regulations led many conservationists to despair. As President Theodore Roosevelt wrote of caribou in 1902, "it would seem the race must become extinct in a comparatively brief period."[35] They were doomed, he believed, but news of their collapse might persuade hunters to slow the slaughter of other species.

Big Bog Caribou: Collapse, Restoration, and Collapse Once Again

Not all conservationists agreed with Roosevelt that woodland caribou would inevitably vanish. Some began to advocate for a new model of conservation: active habitat restoration, followed by translocation of individuals from healthy populations. In the 1930s, wildlife managers decided to attempt a bold experiment, preventing woodland caribou

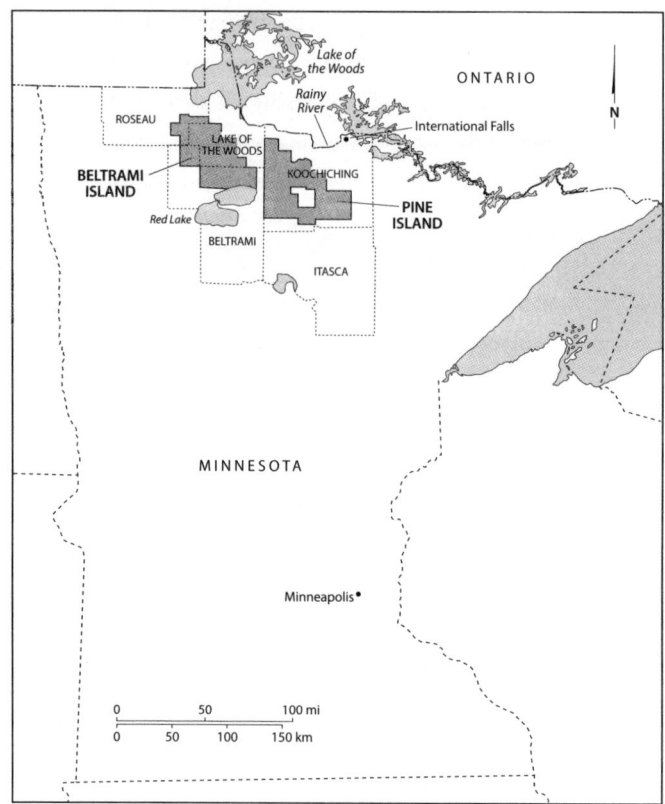

MAP 2.2 The Big Bog Region of Minnesota. *Bill Nelson*

from going extinct in the United States by translocating individuals from Canada.

Long after woodland caribou had retreated from the rest of the United States, they had persisted in northern Minnesota, protected by the vast bog complex known as Big Bog. According to caribou biologists Thomas Bergerud and W. E. Mercer, the herd had likely "wintered on the open bogs of Beltrami, Lake-of-the-Woods, and Koochiching counties, but migrated to the islands of Lake-of-the-Woods and Rainy Lake in Ontario in the spring to avoid predators during calving."[36]

But by the early 1920s, Minnesota caribou were dwindling as well. Homesteaders had drained much of the Big Bog region near Red Lake, hoping to turn the peat wetlands into a rich farmland. These efforts largely failed, as they did across the upper Great Lakes, and the abandoned farmlands reverted to the state. During the Great Depression, managers created a wildlife refuge on the abandoned lands and tried to protect the remaining caribou by restoring their dwindling wetland habitat. Despite all their efforts, the herd continued to decline.

Federal biologists, desperate to save what was then the last herd in the mainland United States, decided to try translocation—always a risky business, but particularly risky with woodland caribou, because no one had ever tried to do it before. After a number of frustrating false starts, biologists finally managed to capture several individuals in Canada and move them to Minnesota, where they penned them inside protective fences, hoping to build a larger herd by sheltering them from predators. Inside the pen, however, the woodland caribou failed to thrive. After some were released, they headed for their traditional migration routes in Canada. But development along the Rainy River blocked their path, and without a way to migrate between calving and wintering grounds, the isolated population failed to sustain itself.[37]

Minnesota's Big Bog, a wetland and forest complex of over 2 million acres, formed ten thousand years ago when Glacial Lake Agassiz burst its ice dams and drained, leaving behind many lakes, including Red Lake, and vast peatlands. Indigenous cultures found—and continue to find—great richness in these wetlands. The Dakota lived there for generations, until they were pushed out by the Red Lake Band of Chippewa in the eighteenth century, who had been themselves pushed west by disruption in the upper Great Lakes.[38]

The Red Lake Nation thrived in the bog and forest mosaics. They developed extensive fishing operations, smoking sturgeon for winter storage (as described in a later chapter), harvesting wild rice from the wetlands, and managing sugar maple forests in the uplands for maple sugar and syrup. Caribou were critical for fresh meat, dried jerky, furs for warm winter clothing and shelter, sinews for sewing, and hosts of other uses.[39]

Even after Euro-Americans formed Minnesota Territory in 1849, pushing most tribal nations off their core lands, the Red Lake Nation defended their ways of life in the wetlands. After complex negotiations with both Dakota and white people, the federal government recognized the Red Lake Nation's claims to the entire northwest corner of what became Minnesota. But in the Treaty of 1863, the Red Lake Nation was forced to cede substantial portions of their territory, while retaining the Big Bog region. As tribal archivist Kathryn Beaulieu writes, "When this treaty was negotiated, the Chippewa Indian leaders were conned into turning over 11 million acres of prime real estate in Northwestern Minnesota and Northeastern North Dakota for about half a million dollars. As far as real estate deals go, the ceding of the Red Lake Valley ranks up there with the Manhattan deal, the Louisiana Purchase, and the Alaska deal. It has been characterized as one of the most dishonest and fraudulent deals ever made."[40]

The Red Lake Nation continued to assert its rights, even while industrial development chewed away at the borders of the reservation. In the 1860s, white-owned logging companies began logging around the reservation and tried to enter reservation forests. Those logging operations threatened both tribal people and the caribou they depended on. Many caribou were shot to supply the logging camps with meat, while others retreated across the Canadian border or into the depths of the bogs. With logging and its resultant fires, white-tailed deer habitat expanded, placing additional pressures on caribou populations.[41]

European homesteaders flocked into the region, hoping to make their fortunes on land taken from the Red Lake Nation. When my husband and I visited the Red Lake Wildlife Management Area in 2019, a sign in the state park visitor's center noted that: "Pictures of cattle standing in tall, green, lush grass" drew immigrants to the bog, pushing tribal members further aside. The next great disruption came with the federal Dawes Act of 1887, which forced many Indigenous nations to allot their common lands to individual families. The privatized land holdings could then be sold to whites. The Red Lake Nation refused allotment, insisting on retaining common lands. In Minnesota, the Nelson Act of 1889 was even more devastating. This law ordered all Anishinaabeg people within Minnesota to relocate to the White Earth Reservation, expropriating the vacated reservations for sale to settlers. Both the Dawes Act and the Nelson Act "called for distribution of Tribal land to individual Indians in hopes of making us farmers who would assimilate into general society." But again the elders of Red Lake refused consent, both to being moved to White Earth and to allotment. The reservation retained—and continues to retain—its status as a closed reservation where all Red Lake members hold land in common.[42]

Nevertheless, settler pressures on the Red Lake reservation intensified. In 1889, the tribe ceded another 3.2 million acres of land in exchange for "promises of money, education, health care, and aid to farming." While a fifty-year trust fund was supposed to be established for the benefit of tribal members, instead "80 percent of Red Lake moneys went into a general fund for all Minnesota Chippewa Indians with only a 14 percent return to Red Lake."[43] In 1902, the Red Lake land base was diminished once again, with 256,132 acres sold to whites.

In 1905, in the midst of these intensifying changes, a white biologist named Charles Johnson made a canoe trip with his brother from Walker Lake to Baudette on Rainy River. His father-in-law, Warren Wood, told Johnson that, until the late 1890s, during the height of industrial logging, caribou were still "common in the territory south of Lake of the Woods, and as far eastward as Red Lake and beyond. In the winter of 1895 he [Wood] had a logging camp on Tamarack river, entering Upper Red Lake on the east shore. The caribou at that time, he tells me, came south in large droves, following the marshes, during the winter." According to some trappers Wood encountered, "at least 500 head went through near their camp." The trappers further reported that "the caribou came down from the north during January and early February, following the muskegs and traveling in bands from 25 to 100 head, with occasional larger bands."[44]

While the woodland caribou that passed through the region had for many generations been migratory, the development of the Great Northern Railroad in the late 1870s severed those migration paths. Johnson reported that, by 1908, some caribou had begun to remain year round in the Red Lake area.[45] This herd had adapted to the intensifying industrial development along the border by changing their ancient migrations, becoming wanderers within a contested territory.

That same year, Minnesota legislators passed the Volstead Act, which sought to drain public "swamplands" in the northern part of the state, particularly those peatlands ceded by the Anishinaabeg. The next year, a land drainage scheme began on peatlands seized from the tribes. Scholar Kristine L. Bradof notes that "plans to drain the peatlands or dam the outlet of Red Lake all failed to consider the wishes of Red Lake leaders."[46]

In Minnesota, the title on lands ceded by the Anishinaabeg were held by the federal government, which decided to sell the land and invest "the proceeds in the commonly held Chippewa in Minnesota Fund." But on lands that weren't sold to homesteaders, or where homesteaders failed to fulfil the terms of the act, the Red Lake Nation retained ownership, even though the lands had technically been ceded. Minnesota state legislators decided to drain those peatlands, in order to "'reclaim' the swamplands, encourage successful homestead settlement, and afford some profit for the Ojibwe from otherwise 'useless' lands. In their eyes, the project would produce revenue for the Indians, revenue for the state, and agricultural land for homesteaders."[47]

From the legislators' perspective, drainage and settled farming was the highest and best use of swamplands. Moreover, they "believed that the bounty they envisioned justified earmarking Ojibwe tribal funds for a percentage of the costs of the land surveys and ditch construction."[48] In other words, the Red Lake Nation would be forced to pay for a drainage project they bitterly opposed, one which would deprive them of access to their lands and waters.

The Red Lake leaders were furious. One scholar writes: "As they articulated their opposition to the drainage project, Red Lake leaders described the different value systems on which they and their opponents based their arguments. The Ojibwe regarded swampland as central to their way of life. The wetlands environment nurtured many of their critical resources. . . . The network of lakes, peatlands, and forest supported the wild rice crop, cranberry bushes, fish, and various animals, especially beaver and muskrat. The entire Ojibwe subsistence cycle depended on the region's water system. Farmers, on the other hand, regarded swampland as useless. Prospective 'developers' described in detail the difficulty of traveling through the muskeg while combating swarms of mosquitos and black flies. They could conceive of these wetlands as having no possible benefit without first 'reclaiming' them to make them suitable for agricultural production. The difference . . . could not have been more profound."[49]

Against tribal objections, the project went forward, with Red Lake tribal funds that had essentially been stolen from the tribe used to finance it.[50] Drainage was an ecological, economic, and cultural disaster—for the Red Lake Nation, and also for the Anglo homesteaders who tried to farm the drained peatlands. Lake levels rose, flooding out tribal fisheries, gardens, and wild rice beds that had sustained tribal members for generations. And peatland water tables were lowered, which most likely encouraged devastating fires in 1910 and 1918.[51]

In 1923, the Minnesota commissioner of immigration, Oscar H. Smith, warned of "unscrupulous land dealers who resort to dishonest and unfair tactics, both in the misrepresentation of lands and in gouging their victims to the limit." Delinquency rates on ditch taxes skyrocketed, reaching 77 percent in 1926 for Beltrami County.[52] But, as late as 1927, state engineers kept encouraging counties to spend tribal monies and tax funds "to buy dredging machines to scour hundreds of miles of choked ditches." Biologists Averell and McGrew complained in 1929 that "very few of the ditches have been maintained, with the result that many of them are partly filled and the channel obstructed by a growth of sedge and rushes in the bottom and brush along the banks. Some ditches had been dammed by beavers."[53]

In 1929, Minnesota's legislature and supreme court finally acknowledged "that the government itself was responsible for pressing Beltrami and other counties into constructing drainage ditches that they could not afford." But rather than return those lands to the Red Lake Nation, the state legislature established the Red Lake Game Preserve on 1.1 million acres of tax-delinquent, former tribal lands—nominally "to protect and to propagate wild life"—but also to prevent counties from defaulting on drainage bonds.[54]

What one biologist called the "far fetched land promotion scheme" devastated caribou and those members of the Red Lake Nation who had continued to rely upon them. By 1928, the total caribou herd in the region had dwindled to only thirteen individuals. Seven years later, in 1935, the Beltrami Island Resettlement Project began, relocating settlers, ending poaching, blocking drainage ditches, and creating wallows for wildlife. Staff built dams "to raise the water level on 200,000 acres; wallows have been dynamited to afford caribou an escape from insects; a four square mile enclosure has been fenced for better control and protection."[55]

But trying to protect caribou meant first figuring out where they were and how many were left. Counting caribou in a bog isn't easy. In 1935, biologist W. J. Breckenridge, with the Minnesota Museum of Natural History, recounted his continuing efforts to survey, understand, and manage the Big Bog caribou herd. Breckenridge called them "this band of survivors" and noted how elusive they were, how difficult to census, track, and know in scientific, quantitative detail. The herd, he wrote, had been frequently "estimated but never accurately determined." With several other state and federal biologists, Breckenridge spent a cold winter week on snowshoes tracking the herd. For days, he found little sign, fearing that they had vanished for good, like ghosts. He wrote: "it may be that these caribou survive only in myths." But the crew persisted, stumbling across drainage ditches in their snowshoes, "with eager hopes of seeing . . . this last rumored herd of a dozen or so caribou which, if really found, will constitute the final survivors of their race in the United States."[56]

Breckenridge loathed the ditches, calling them first a monster, then an octopus. He wrote: "Crunch—crunch—crunch, continues from the bed of the old drainage ditch, the thoroughfare for winter travel in the muskeg . . . the monster of the region which has gulped up hundreds of thousands of dollars, has dried up the muskeg into tinder for brush fires, has produced little or no fertile lands for tillage, and now lies like a giant octopus with its tentacles grasping the land and slowly but surely rendering it of use to neither man nor animal."[57] Caribou stumbled into ditches and could not escape, and the carcasses of the drowned and frozen animals saddened the men. But his mood lifted when, after searching for days, they finally found tracks of caribou in the snow and knew "that the caribou, although surely miles away, must really exist somewhere in this vast muskeg."[58]

In 1936, biologist Gustav Swanson joined the restoration team, in part to galvanize the political will among wildlife biologists necessary for a risky translocation program. Swanson argued that caribou extinction was tragic in and of itself, but its importance could only be understood in the context of global extinction history. He wrote: "No conservationist's plea for action is complete until he has recited the tragic story of the passenger pigeon, the great auk, the Labrador duck, the Eskimo curlew, the California condor, and the heath hen. . . .

My purpose is to bring to your attention just one more species which we may properly add to this list"[59]—unless of course, the state and federal government took bold action to stem the tide of extinction.

Yet woodland caribou were different, Swanson argued, because they were not yet on the verge of global extinction, only extinction within the United States. That made them worth the extraordinary efforts needed to restore them. If the nation could keep one ghost population from being driven to extinction, those ghosts would return to flesh and blood and inspire other restoration efforts. He argued that while most wildlife was threatened because of habitat loss, the caribou didn't lack breeding habitat, because they lived on "a portion of the Red Lake Game Refuge in northern Minnesota. The enormous bog offered caribou refuge from human harassment." Yet even with this substantial protected area, the herd had shrunk to a mere eight individuals. Swanson asked: why had hunting regulations and habitat protection "utterly failed in the case of the woodland caribou? The protection which has been extended for several decades has not succeeded in checking the decline."[60]

The fundamental problem, Swanson felt, was migration: caribou were wanderers, and once they left the protected area, they were vulnerable. Swanson worried that "occasionally they have been known to wander many miles in any direction from the big bog in which they are safest, and at these times they expose themselves to the danger of succumbing to a frontiersman's appetite for fresh meat." Swanson also noted that wolf populations had grown inside the refuge. Typically wolves have ecological value and should not be harmed, Swanson wrote, adding "I would be the last to encourage their persecution."[61] But because drainage had offered wolves easy access to caribou, he thought it necessary to protect the remaining herd from wolves until caribou numbers had rebounded. And so they hired a warden to trap wolves out of the caribou calving range.[62]

Swanson feared that because tribal lands bordered the caribou range, Red Lake Nation members would inevitably kill caribou. So he insisted on a fence between the refuge and the reservation, one long enough to keep Native Americans and woodland caribou from encountering each other. This was, perhaps, the fatal flaw of the project, for it severed the caribou from any opportunity of migration.

Swanson had recognized that migration was part of the caribou's evolutionary history, and he had argued that a loss of migration corridors was the ultimate cause of the caribou's decline. Yet he still advocated for a massive fence to keep tribal members away. Why? He feared poaching, telling his fellow wildlife biologists: "When it is pointed out, furthermore, that a large Indian reservation directly adjoins the game refuge the danger from illegal shooting which the caribou face is more easily realized. When the caribou wander a scant dozen miles southwest of their usual range they are in the heart of the Indian hunting grounds, and they have been found just there within the past 2 years." Swanson went on to admit "we have no definite records of caribou taken by hunters legally or illegally in the past several years, but the territory in question is so difficult to patrol that detection of law violations is certainly the exception rather than the rule." He felt that "to save the animals from wandering into the Indian reservation where they would be constant danger," the only solution would be a fence 16 to 22 miles long—a massive undertaking.[63]

Like other Euro-American wildlife biologists at the time, Swanson could not imagine that Native Americans might be able to regulate their own hunting. Swanson was trained in a logic of conservation that saw both wolf predation and Indigenous hunting as rapacious and uncontrolled. The tragedy is that, while he could recognize the flaws in a blind hatred of predators, saying that "I would be the last to encourage [wolf] persecution," he could not recognize the racism that structured his own recommendations.

Swanson's paper was delivered at a wildlife society meeting, and the comments it received are fascinating. One wildlife biologist declared that they could not possibly "have coyotes and wolves on same ground [as the caribou]. Takes money, but they can be cleaned out."[64] As a profession, wildlife biology in the 1930s continued to extend the same logic to Indigenous peoples, with conservationists believing that elimination of tribal hunting was critical for restoration success.

1938 Efforts at Translocation

In 1938 the federal government—represented by the New Deal's Farm Security Administration and Soil Conservation Service—decided to move forward on a translocation project to supplement the dwindling herd. The 1938 reports of trying to catch caribou from Saskatchewan are noteworthy in their portrayal of both the uncertainty and the hard physical labor involved in this project. Biologists, working with Indigenous hunters from Canadian First Nations, struggled to make their way through the snow and trap caribou without harming them. They tried to reduce mortality while calves died of diarrhea or infection, adults broke their bones struggling to escape from the traps, and translocated females simply wasted away in their new Minnesota home. The written records of the biologists make it clear that they cared deeply for the suffering and fate of those caribou as individuals.[65] They didn't want to needlessly inflict pain on the animals they captured. And they believed in the project's vision: woodland caribou restored to the Big Bog, a reversal of the devastation wrought by a vision of human progress that insisted farmers could overcome nature's constraints and reshape the face of the bogs.[66]

John Manweiler, project game manager with the Soil Conservation Service, worked with the Big Bog caribou restoration project for years and wrote numerous internal reports and public articles. He started a 1938 public piece in the magazine *Parks and Recreation* with a quote from soil scientist Walter Lowdermilk: "Many ancient civilizations, once revelling in a golden age of prosperity, are crumbling in ruins or lie buried in sands and debris, largely caused by the destructive treatment of the lands on which they were dependent for sustenance. After a long struggle, a civilization either died or its people migrated to more productive regions."[67] Like other New Deal conservationists, Manweiler was inspired by the hope that people could write a new narrative of environmental history, one of restoration rather than destruction.

In another article, Manweiler detailed why the Farm Security Administration was spending such a lot of time and money on woodland caribou rather than farms. All land should be put to its highest productive use, he agreed—but a great deal of land was actually submarginal for farming. Instead, a rational vision of production would make the land produce what it was best suited for: "developing it for the production of forestry and wildlife, on a large scale."[68] In some ways, this is a vision of centralized planning, a hope that rational mapping would allow land economists and land scientists to maximize production and utility. It's easy to sneer at such a vision. But unlike the plans that James Scott critiques throughout *Seeing Like a State*, this was not a vision developed through a distant synopticon, far above the surface of the bog.[69] Rather, people waded through muskeg, slapped at mosquitoes, fell awkwardly on their knees in their hip waders while attempting to admire pitcher plants and orchids, and shaped their vision of restoration into a workable plan. A history of loss and a fear of past extinctions shadowed their efforts, motivating them to move beyond despair into action.

Manweiler recognized that caribou decline had been shaped by loss of migration routes and restriction to a small range.[70] Calf predation, disease, inbreeding, and the "drowning of young in drainage ditches" left behind by failed farmers added to the problem. He wrote: "After two years of study it was realized that the animals had passed the minimum survival point and that if any of the animals were to be saved, new stock must be introduced." Managers decided to enclose four miles of pasture for the caribou, fearing they would wander out of the bog and be poached by tribal members on the reservation.[71]

Manweiler acknowledged that penning the caribou had a certain irony. "The animals used to migrate freely between American and Canadian territory in the vicinity of Lake of the Woods," he wrote, "but with the advent of settlement of the fertile lands bordering that lake and the Rainy river, the Minnesota caribou became separated from the Canadian herds and were virtually trapped in the huge bog area." But, he added, experimentation was necessary to avert "complete disappearance of this valuable big game species. . . . . Suffice it to say that one of the most spectacular experiments in the history of game management has been carried to a successful completion."[72]

William Cox, head of the project, was equally hopeful in an article published in 1939. Like the others, he began his essay by framing their restoration efforts within the historic context of extinction. Cox wrote: "Not so many years ago heath hens abounded in the New England States, and passenger pigeons were so numerous in the Mississippi Valley that their flight sometimes completely obscured the light of the sun. Today, both species of birds are extinct—destroyed by a wave of remorseless slaughter that will remain always a blot on the sportsmanship of the white man. But until a few years ago, not many people were aware that a similar fate imminently threatened one of the finest species of big game in the United States: the woodland caribou (*Rangifer caribou-sylvestris*). . . . The species had in fact almost reached the vanishing point in the United States before anything was done to save it."[73]

Cox noted that drainage ditches were another historic legacy of failed farming. He called the ditches "veritable traps for big game . . . so soft on the bottom that large animals were mired in them, and their banks were steep and high. Rangers often found dead animals in them." The ditches, he feared, also changed predation, favoring wolves over caribou. "Before drainage work was started the cows could find little islands far out in the wet swamp where they could drop their calves comparatively free from molestation by wolves. After the swamps were drained, the wolves could pick up the calves readily without the necessity of swimming or wading to the islands."[74]

Despite these efforts, by 1941 caribou had become harder and harder to find in the pastures.[75] By 1947, no sign at all remained of the translocated caribou.[76] This is a sad story, particularly in the slow violence of its ending. In 1965, an analysis of the project by biologist Bernard Fashingbauer both recognized its failures and celebrated the attempt. Fashingbauer noted that "Relatively little attention has been given to management of woodland caribou because the animal was already gone or had suffered serious declines in the more accessible portions of its range before the advent of modern wildlife management."[77] Because few biologists considered caribou to have any chance of persistence, they had largely been forgotten. He urged Minnesota's wildlife agency to keep trying to restore caribou, even though the Big Bog effort had not sustained a permanent herd. "A glimpse of woodland caribou restored to their native habitat could provide great pleasure to the wilderness traveler in Minnesota," he argued. "Time, study, and public interest can make this a reality."[78]

In many ways, the Big Bog project was a bold attempt at a new restoration paradigm. Managers built on lessons learned from past extinctions, and they hoped to create not just a new breeding population of woodland caribou, but a new ethic of wildlife management focused on more than just the utilitarian regulation of hunting.

But for all its idealism, the project failed for two main reasons: first, migration corridors remained severed. Without the ability to roam and escape predation, caribou withered away. But even more critical was the project's failure to address Anishinaabeg rights and relations. Caribou loss was part of a much larger context of Indigenous trauma, and by ignoring that—indeed, by perpetuating the racism that underlay that trauma—project managers doomed the project to failure. Euro-American wildlife managers were determined to reverse the effects of development, with its impulse toward industrial expansion. Yet because those managers were largely blind to the racism against Indigenous peoples that provided the basis for Anglo expansion, they replicated the same ecological problems they were trying to ameliorate.

THREE

Caribou Futures
in a Warming World

O n the Canadian side of Lake Superior, the population of woodland caribou persisted longer than in the United States. Yet they too have "been driven off the land," according to Leo Lepiano, former Lands and Resources Consultation Coordinator for the Michipicoten First Nation. By 1912, the Lake Superior woodland caribou had been hunted out from the western shore and Thunder Bay region. In the Lake Nipigon watershed just north of Lake Superior, caribou thrived until the Canadian National Railway came through in 1910, when that population also began to dwindle (although islands in Nipigon Lake may still support some breeding woodland caribou).[1]

Along the north-central and northeastern shores of Lake Superior, the range of woodland caribou remained continuous all the way north to Hudson Bay well into the mid-twentieth century. After World War II, however, mineral, forest, and energy development altered and fragmented their range, and caribou populations became discontinuous, with the Lake Superior population cut off from the more northern populations. This wasn't a northwards migration, with caribou expanding into new territory. Retired ministry biologist Gordon Eason described it as a cascade of local extinctions driven by hunting, predation, and habitat loss. From 1880 to 1990, the woodland caribou extirpation in Ontario moved northward at about 20 miles per decade.[2]

Well into the twentieth century, Lake Superior woodland caribou had seemed to be persisting along the north central and northeast shore of Lake Superior, particularly in Pukaskwa National Park and on islands that were free from wolves such as the Slate Islands, 12 kilometers offshore. When wolves were absent or in low numbers, woodland caribou populations were able to increase sharply, even to the point of denuding forest habitat on small islands. Canadian biologists were eager to restore caribou to sites they had recently occupied, and the Slate Islands were an opportune place to capture caribou for translocations to suitable habitats elsewhere.

Beginning in 1982, the Ontario Ministry of Natural Resources translocated woodland caribou from the Slate Islands to a number of other islands, as well as to the Lake Superior shoreline. Biologist Gordon Eason was involved with most of these efforts, and even in retirement he remains deeply involved with translocations of woodland caribou in the Lake Superior region, working hard to keep the population from extinction. In several visits and interviews over ten years (two conversations in the summer of 2010; one long conversation and lunch in the summer of 2018; and a visit over two days in February 2020), Eason described to me their attempt to stem the decline of woodland caribou by translocating individuals in the eastern part of Lake Superior.

The island sites they chose were free of white-tailed deer and large predators, while the mainland site was deemed large enough to give caribou a chance to escape wolves, bear, and lynx. In 1982, they moved eight individuals from the Slate Islands to Michipicoten Island. A single adult male was already on Michipicoten, one that had presumably walked over the ice from the mainland near Pukaskwa National Park. Those caribou thrived (until wolves arrived), even though Michipicoten Island is mainly Great Lakes forest with little boreal forest or ground lichen.

In 1984 and 1986, caribou were moved to small islands just off Lake Superior Provincial Park, close enough to the mainland so that individuals could migrate back and forth, but far enough to provide some protection from wolves during calving. The Montreal Island population was started with eight caribou from the Slates. Two caribou were successfully moved to Leach Island to supplement a lone female who had moved there from Montreal Island. The lone bull left after two years leaving a population of four females.[3]

What biologists Gagon and Cochran call "the most ambitious effort to establish woodland caribou along the Lake Superior shoreline" began in fall 1989 on the Lake Superior Provincial Park mainland, near the spectacular Gargantua Peninsula. When I first kayaked there on an overnight camping trip in 2003, we listened to our guide tell stories about these caribou and the ancient pits from the first peoples who followed the ice north and hunted in this spectacular landscape. While wolves are common in the park—and we imagined we heard their howls that night—the rugged terrain and many nearshore islands offer escape for the caribou. The project eventually moved seventeen radio-collared caribou and twenty-two uncollared caribou to the Gargantua Peninsula and two offshore islands.[4] The population persisted for two decades but, by 2010, no sign of their presence remained.

That year, I spent a week sea kayaking and camping along this rough coastline, hoping to catch a glimpse of a caribou, to no avail. We explored Pukaskwa pits, stone-lined depressions left by ancestors of the Anishinaabeg many centuries ago. Most archaeologists now believe these pits may have been food caches; caribou teeth, charred caribou bones, and flints have all been excavated from these pits.[5] We spent one powerful afternoon paddling by the Agawa pictographs, a sacred Anishinaabeg site on the coast. Here, rock paintings of caribou, sturgeon, and loons (among many other animals, including a monstrous lynx figure called Mishipeshu), speak to the caribou's religious importance for the region's Indigenous peoples.[6] For many thousands of years, caribou had vital material as well as spiritual value for the First Nations peoples in the watershed—yet trying to keep caribou on the land has been a struggle in the decades since settler colonists pushed Indigenous people off their lands and waters.

Eventually, the Montreal Island, Leach Island, and Michipicoten Island caribou failed to persist as well. The repeated failure of efforts to translocate caribou and help them escape from predation suggest that the threats facing the population are interconnected. Predation and brainworm parasitism are the proximate stressors leading to caribou decline. But these stressors occur in the context of larger ecological changes: habitat alteration and fragmentation caused by industrial forestry and energy infrastructure such as powerlines, roads, and pipelines—along with a warming climate. All of these changes create conditions that favor moose, and the increased biomass of prey offered by moose leads to increased wolf populations. Wolves then prey selectively on the caribou, which are smaller than moose and easier targets. Habitat alterations also favor white-tailed deer, hosts for brainworm, making caribou more vulnerable to parasitism.[7]

For woodland caribou to persist in the longer term, reducing the density of roads, trails, and snowmobile paths is necessary across the boreal forest. Winter travel routes that are convenient for people are also convenient for wolves. These human-made paths facilitate encounters between wolves and caribou by allowing wolves to cover more area. When anthropogenic paths give wolves an advantage, caribou evolutionary strategies honed over millennia no longer work.

When wolves were absent on the Slate Islands and Michipicoten Island, caribou herds flourished for decades. At 184 square kilometers, Michipicoten Island was large enough to support a substantial caribou herd. By 2014, the population on Michipicoten had grown exponentially, from the initial 9 to perhaps 920 individuals. On the Slate Islands, caribou populations had also increased.

But then came wolves. In the cold winter of 2013–2014, ice bridges formed to both the Slate and Michipicoten Islands. Several wolves took advantage of those bridges and walked across. With abundant food in the form of caribou, those few wolves soon had pups, then more pups. On the Slate Islands, caribou dwindled after wolves arrived, until the wolves starved or left the islands, their prey base gone. On Michipicoten Island, four wolves rapidly multiplied and the caribou were unable to find refuge.

Caribou in the Lake Superior Basin: What Lies in Wait for Them?

Soon after wolves appeared on Michipicoten Island, an informal coalition of the Michipicoten First Nations Community, local cottagers (including Christian Schroeder), and caribou biologists urged the Ministry of Natural Resources and Forestry to protect the Lake Superior caribou from extinction. Options discussed included culling the island's wolves (unpopular with wolf advocates) or moving wolves to Isle Royale in Michigan, where the U.S. National Park Service had for years been considering a wolf restoration effort. Alternatively, if wolf removal was impossible, the only choice was to rescue some caribou to wolf-free islands.

The ministry, however, initially refused to act, arguing that watching predation play out would be an interesting scientific experiment. "We're still thinking of allowing nature to take its course," one of the ministry staff explained to Lepiano.

A later decision to translocate caribou to the Slates archipelago and Caribou Islands has met with the general approval of the informal community coalition. But group members remain concerned. In a 2018 interview, Lepiano and Schroeder discussed the continuing political and environmental forces affecting the caribou. Lepiano noted that if the caribou vanished, many management dilemmas faced by industry and the ministry might vanish as well. "There's a tremendous political pressure from certain interest groups to not have to deal with the caribou on the mainland," Lepiano said. Currently, development in Canadian woodland caribou range cannot exceed 35 percent of the landscape, as outlined in the 2012 federal recovery strategy.[8] That limit would be lifted for the so-called "discontinuous range" between Lake Superior and the northern populations if Lake Superior caribou went extinct. Schroeder, an advocate for translocation, argued that more of the boreal forest might then be opened to logging disturbance, accompanied by the expansion of a huge mining project known as the Ring of Fire. Schroeder said: "The transmission line that they're running through here is intended to run power to the mining infrastructure in the Ring of Fire, which ironically I will remind you, is prime caribou habitat in northwestern Ontario. And caribou are a real pain in the butt."

Schroeder suggested that many biologists and policymakers within the ministry share a perception with environmental NGOs that climate change will inevitably doom the caribou. "Caribou in Canada may be doomed by climate change and habitat loss" proclaimed one headline in *Nature World News* on December 16, 2013. A 2017 paper by Sara Masood and colleagues projected "complete loss of woodland caribou in Ontario if winter temperatures increase by more than 5.6 [degrees Celsius] by 2070."[9] As one caribou biologist within the ministry said, according to Schroeder, "climate change is going to wipe caribou out anyway," so why expend resources to save them, when those resources could be used on other, less vulnerable species?

What's next for the Lake Superior woodland caribou? The ministry has solicited advice from the public on the development of a revised management approach for the Lake Superior population, and the staff planned to hold Indigenous community and stakeholder consultations in due time. Possibilities range from doing nothing, to maintaining caribou on offshore islands, to trying to restore connectivity to northern populations.

What Now?

Across Canada, similar dilemmas are confronting wildlife managers and citizens. When protecting caribou means killing wolves, what should we do, now that most urban environmentalists have learned to love wolves? Wolves were once loathed by settler colonists in North America with the passion of a thousand fires.[10] After fierce eradication efforts—many in the name of game conservation—led to their near-extinction in the Lower 48, wolves returned when federal Endangered Species Act protections went into place.

Aldo Leopold's powerful essay, "Thinking Like a Mountain," best exemplifies this shift. In it, he writes about how, as a young man working in wildlife, he believed that killing wolves was always a good thing. Fewer wolves surely meant more game. But the day he shot a wolf and watched the fierce green fire fade from her eyes, he began to "think like a mountain," eventually articulating an ecological perspective on predators and prey.[11]

Among many urban and suburban folks, a loathing of wolves has changed to love for wolves. Wolves represent wilderness, the hope of restoration, the promise of nature unbound, uncontrolled by humans. I understand this. When I visited the wolf restoration site at Yellowstone in the summer of 1995, I was thrilled when I caught a glimpse of a restored pack. Several years later, a group of environmental historians were lucky enough to tour winter Yellowstone with biologist Doug Smith, one of the leaders of the restoration project, and we spent hours stomping our frozen feet at spotting scopes, watching wolves romp. For twenty-five years, I have taught wolf restoration as a classic case study in my environmental studies classes. I have a picture of a howling wolf above my desk.

And yet. Making wolves into untouchable emblems of wilderness won't ultimately protect them on the landscape. In the fall of 2019, I was drinking with friends around a bar's firepit in Burns, Oregon, a small ranching town. (Remember bars? Remember hanging out with strangers?) My friends and I got to talking with a couple of nice guys, who I will call Jack and John, computer programmers living in Boise, Idaho. They had grown up on ranches and still liked to fish and hunt, but they also liked their suburban houses and steady jobs. We were shooting the breeze, enjoying the cool evening and the smell of juniper burning on the fire.

Then they started talking about wolves in Idaho.

"Those wolves hunt in packs of over a hundred!" Jack insisted.

"A hundred? A hundred wolves in a single pack?" I asked.

"Yeah! That's right! A hundred wolves, hundred and fifty wolves, all hunting in one pack!"

John chimed in, "And you should see them! Those wolves are huge! They can get to be 200 pounds in size! One of those wolves alone can take down an entire herd of elk!"

"Really?" I asked, and the guys whipped out their phones, where they had what they saw as proof—an image of a really big dead wolf in some stranger's arms.

"And they're coming into the suburbs and snatching dogs out of the yard!" Both John and Jack agreed on this.

Of course the photo was manipulated, and of course the rumors of wolves hunting in massive packs are imaginary. Nonetheless, John and Jack were convinced that massive wolves and massive wolf packs were taking over Idaho, threatening wildlife, and possibly even their suburban families.

Remember the nineteenth century, when white settlers shivered in their sod houses on the Great Plains, terrifying themselves with wolf stories? Willa Cather in *My Antonia* retells a particularly terrifying tale: a massive pack of wolves in Russia slaughtered dozens of people riding home in their sleighs from a wedding, killing all but two people—who found themselves banished to America. When I read this story aloud to my class, with the lights dimmed, it never fails to make the students shiver. We still haven't forgotten our fear of wolves, no matter how much we love them on our computer screens.

Ironically, the best way to ultimately protect wolf populations may be to allow the killing of some individuals. Otherwise, Euro-American fears of wolves may sweep aside reason, and rational discourse may turn into terror. But the ethical ambiguities of this position trouble many.[12]

When the caribou coalition from Wawa suggested to the superintendent of Pukaskwa National Park an experiment in caribou translocation to create a new mainland population, the superintendent refused to consider the proposal until a much longer process had been undertaken. Before considering translocation ideas, she wanted to see a larger stakeholder survey, a habitat plan that addressed the long-term persistence of caribou in the park without any human intervention, and a study of the ethics of translocation into ancestral habitat if that habitat had become a sink for caribou. These are all excellent topics for study. But they would require so much time to address that the population in question would be extirpated long before the studies were completed. In addition, these concerns share the premise that a translocation is successful only if the population can survive in perpetuity without human intervention. That premise would doom most rare species on earth to extinction.

Across Canada, wildlife biologists are taking increasingly desperate measures to keep the remaining herds from crashing.[13] But when these measures include wolf culls, they generate controversy. In the winter of 2019–2020, a total of 463 wolves were killed—an action justified by a study led by eminent caribou biologist Robert Serrouya. Serrouya and colleagues had argued that penning pregnant female caribou and killing wolves—in the short term—was significantly more effective than restoring habitat damaged by energy extraction and industrial forestry. Serrouya told a journalist that habitat restoration would be "too slow" because "herds don't have the decades that takes."[14] Or as the formal scientific paper he coauthored argues: "The continental scale of forest alteration and extended time needed for forest recovery means that relying only on habitat protection and restoration will likely fail. Therefore, population management is also needed as an emergency measure to avoid further extirpation."[15]

Other scientists found fault in the paper's analysis.[16] In *The Atlantic Monthly*, journalist Sabrina Imbler writes that "killing wolves is often controversial, and in this case their deaths may have been in vain: A group of scientists says the decision to cull the wolves rested on a statistical error."[17] However consequential this statistical error may have been, it doesn't change the fact that nearly all caribou biologists agree that, for caribou restoration to succeed, it's essential to protect calving caribou from predation. There are different ways to achieve this: kill 80 percent of the wolves in the area; move caribou away from the wolves; pen pregnant cows. Or one could consider the strategy developed over millennia by reindeer peoples in the north: follow the caribou on their migrations.

As the Lake Superior example shows, protection from predation may be necessary to stem the crash in many woodland caribou populations. But it only buys time for a broader adaptive management strategy. Mark Hepplewhite, a caribou biologist who is one of the coauthors of the Serrouya paper, clarifies that culls alone don't work. Hepplewhite says "in order to have any effect, 50 to 70 percent of the wolf population that feeds on the herds must be killed through government-sponsored aerial or ground shooting programs." He admitted to his interviewer, however, that the impact on caribou populations can be minimal. "Alberta has probably killed on average 150 wolves a year in one small caribou range of 80 to 100 caribou. It actually was ineffective at growing the caribou population. All it did was prevent the decline."[18]

But staff with many provincial governments seem happy to rely on wolf culls, interpreting studies like that of Serrouya and his colleagues to mean they can essentially ignore habitat issues. In 2018, Quebec washed its hands of another genetically distinct caribou population, saying that trying to prevent them from extinction was a waste of money. As the Canadian Broadcasting Corporation reported, "Forests, Wildlife and Parks Minister Luc Blanchette said the province will not pay the $76 million that his ministry estimates is necessary over the next 50 years to protect the habitat of the Val-d'Or caribou, only found in Quebec's Abitibi-Témiscamingue. In a report published this week, the ministry estimates there are only 18 of these animals still roaming the forests south of Val-d'Or. 'It's a sad situation, but we have to be reasonable,' the minister said. He then added: 'blocking economic activities on the land occupied by the caribou could cost the region up to 187 direct and indirect jobs.'"[19]

In 1989 the province had designated 2160 square kilometers of forest as a protected zone for these specific caribou. But less than half of that protected zone was closed off from clearcutting and mining—and so caribou continued their steep decline.

In Alberta, caribou ranges overlap oil and gas leases. The provincial government has continued to grant these leases for tar sands development, even though provinces are expected by the federal government to develop policies that sustain caribou, rather than lead to their extinction. In Alberta, caribou populations in ranges that overlap tar sands are shrinking rapidly.[20] However, Alberta, like Quebec, has refused to list woodland caribou populations as endangered, because that might mean habitat protection plans would need to be drawn up, thus limiting extractive energy development.[21]

The Conservative government of British Columbia took the Serrouya et al. study as justification for ending restoration projects, not as a way to buy some time to allow habitat protections for caribou to work. In September 2019, the government decided not to "designate any new caribou-protection areas for the deep-snow caribou." According to Imbler, "This spelled disaster for the caribou, which have experienced one of the steepest population declines of any caribou population in the world, as their forest habitat has been razed by clear-cutting or fragmented by roads. The deep-snow caribou once roamed as far south as Montana's Bitterroot Mountains, but almost all populations had disappeared from the United States by 1980. By the early 2000s in Canada, the deep-snow caribou had lost as much as 45 percent of their population in just 27 years. But the government had an alternate plan: a new wolf-cull program that cited Serrouya's study as proof of why killing wolves works."[22]

Climate Change in the Boreal Forest and Caribou Futures

Woodland caribou are largely creatures of the boreal forest, and so the fate of the boreal forest in the Anthropocene will also influence the fate of woodland caribou. The boreal forest makes up one of the world's largest terrestrial ecosystems, extending for more than 16 million square kilometers across North America, Europe, and Russia. In recent decades, tracts of boreal forest have become the target of commercial forestry operations, eager to find new sources of pulp to satisfy the world's growing appetite for paper. And now they are becoming key sites of conflict in the twenty-first century, as the rush to exploit tar sands and other energy deposits within boreal forests crashes into growing concerns about climate change.[23]

Boreal forests are disturbance-prone ecosystems, which means that fire, wind storms, and insect epidemics have been part of their ecological history. Climate change, however, means that the scales of these disturbances are changing. Fire intensities are increasing, insect epidemics are intensifying, and toxic substances are saturating the aquatic ecosystems that make up such an important part of boreal landscapes.

Climate change is also transforming the ecological relationships that tie together boreal ecosystems. As increasing temperatures reduce the frequency of late spring frosts, for example, budworms may have more time to reproduce. The changing climate also appears to be decoupling budworm population cycles from those of its predators, both parasites and birds. Three warbler species that feed on budworms may be shifting their ranges north at faster rates than are the budworms, making the birds less likely to control outbreaks.[24]

Forests will change across the north, with models predicting that the north woods of Wisconsin and Michigan may become similar to forests that now exist in Arkansas. By the end of the century, the cool summer climate of Michigan's Upper Peninsula will be more like the current climate of southwestern Kansas.[25]

Nearly 85 percent of the Lake Superior basin is currently forested, with a mixture of boreal forests in the north and aspen-birch, spruce-fir, maple-yellow birch, and white-red-jack pines along the southern shores. Climate change means that existing stresses on trees will increase, including drought, wind, fires, and insects, leading to large-scale tree mortality and reduced regeneration of trees from increased deer herbivory. With all that dead wood, forest fires may increase. Summer drought means that boreal forests along the north shore of Lake Superior may vanish if continued warmer air degrades habitat for cold-loving trees like paper birch, white spruce, balsam fir, and white cedar (see plate 3).

Rising temperatures, drought conditions, and insect damage is likely to increase fire frequency and intensity in boreal forests. In the short-term, stand-replacing fires destroy the ground-lichens that caribou eat in the winter. Historically, fire has served as an essential disturbance in boreal forests. But with rising temperatures, many fire ecologists fear that increasingly intense and hot fires may lead to major ecosystem shifts in caribou habitat: namely, a shift to grassland or grass/shrubland in areas currently dominated by the southern boreal

forests. Current research suggests that boreal forest in Saskatchewan may well be replaced by grassland vegetation by the end of this
century.[26]

In June 2003, Canada's Standing Senate Committee on Agriculture
and Forestry called for action to conserve the nation's boreal forest.
Some protections followed. But those have been largely focused on the
northernmost reaches of the forest, where logging has less commercial
viability. The southern boreal forest—where the highest diversity of
birds and other wildlife are found—is mostly still slated for logging.
Commercial ventures are moving quickly to utilize timber in the
southern boreal. As one report warns: "More than 30% of Canada's
boreal forest has already been allocated for industrial development,
most of it since 1990. . . . Technological advances are making previously noncommercial forests accessible for harvest; and the search
for oil, gas, minerals and hydro-electric power is moving further into
once-remote territory."[27]

A 2009 study found that just under 2 million acres of boreal forest
in Canada were harvested each year, with 65 percent of that going
to pulp and paper production—often for low-value products such as
advertisements and toilet paper.[28] Some paper companies are quite
explicit about their goals for the boreal forest: get what's left before
it's gone. Why let perfectly good fiber go to waste, especially if the
forest is going to die anyway as the climate warms? Some government
agencies have adopted a similar attitude: it makes little sense to waste
resources on woodland caribou if climate change is going to doom
their boreal forest habitat.

But what climate change means for the intricate relationships between woodland caribou, wolves, and white-tailed deer isn't simple or
straightforward. In the short term, over the next half-century, snowfall
will likely increase in much of the region—even as temperatures
get warmer. This will occur because ice cover on Lake Superior is
decreasing, and less ice cover means more lake-effect snow. Because

woodland caribou evolved in a world of deep snow, this snow may help them survive pressure from predators, while also keeping white-tailed deer at bay.

But as winter weather continues to moderate, some winter precipitation eventually may fall as rain rather than snow. This is already happening, in short bouts during many winters. When that rain freezes on the ground, the resulting ice means that caribou have a very hard time scraping down to their winter foods. Although they can dig through 5 feet of snow to reach ground lichen, a thin sheet of ice can mean starvation. Deer, on the other hand, do much better in warmer, wetter winters, and so deer populations may gradually increase—and caribou will struggle with the brainworm parasite carried by deer hosts.

Warming waters are already reducing winter ice cover on Lake Superior, which also affects predation risk for caribou. Observed ice cover on Lake Superior has dropped 79 percent in recent decades—2 percent per year for the period from 1973 to 2010. In February, the lake used to average 75 percent ice cover but, by the end of this century, we'll be lucky to get 10 percent ice cover. Ice duration is also a lot shorter: it may drop by 60 days—to less than half its current duration—by century's end.[29] When ice bridges stop forming, caribou may find it harder to cross to their island refugia—but their wolf predators may also find it harder. The patterns of movement will change, that's for certain, and for a species as movement-oriented as caribou, that may have profound—but uncertain—effects.

Climate change threatens boreal forests across the globe. But does that mean we should give up on woodland caribou? Or has climate change become an excuse to avoid the difficult political choices that might restore woodland caribou? Given half a chance, caribou may be far less vulnerable to climate change than many other northern species. Unlike moose that have trouble foraging when temperatures warm, woodland caribou don't experience thermal stress—at least not in the range of temperatures predicted for the Lake Superior region.

Moose begin to experience physiological stress at 14 degrees Celsius, panting and foraging less. Caribou, however, don't show measurable physiological responses until temperatures reach 35 degrees Celsius.

Popular perception holds that wintering woodland caribou require old-growth boreal forest with abundant lichens—and those forest types probably won't survive along much of Lake Superior as the climate warms. But does that mean caribou also cannot survive? Not at all. While caribou select lichens in winter if they're available, they can adapt to other foods when necessary. Michipicoten Island, for example, has a mixed forest dominated by hardwoods, and winter lichens are rare. Yet caribou thrive there in the absence of wolves.

Caribou responses to historical climate change offer clues to how caribou might respond now. As mentioned earlier, caribou have survived repeated glaciations over the millennia by moving to ice-free refugia. At the end of the last glacial maximum, the fossil record shows that caribou vanished from those regions as they warmed. Anthropologists Don Grayson and Francoise Delpech interpret this bit of history as evidence of the profound vulnerability of caribou to future climate change.[30] But this interpretation reduces the agency of caribou themselves. Post-glacial caribou didn't simply go extinct. Rather, they chased the melting ice north, exploring new environments that opened up as the climate warmed and expanding their range across the circumpolar north. Migration was central to the caribou's post-Pleistocene resiliency, suggesting that they can be resilient if their potential habitats are connected. Each time the ice retreated in an interglacial period, caribou followed the melting ice north, expanding into new habitats across a diverse, warming landscape.[31]

Caribou Created Our World

Why should we care about caribou histories in an uncertain future? In 2003, biologist Valerius Geist argued that "broad public support

and determined effort by volunteers is essential for wildlife conservation." Geist urged us to consider the deep interconnections between
human and caribou histories. "Reindeer feature prominently in the
rise of modern humans and the demise of Neanderthal man early in
the Upper Paleolithic. The colonization by humans of the periglacial
environments during the last glaciation depended on the rich periglacial megafauna, *Rangifer* included. . . . Modern humans owe much
of what they are to reindeer. We need to reciprocate."[32]

People followed the caribou north in that warming world, and a
striking diversity of human-caribou relationships developed. According to anthropologist Piers Vitebsky, caribou made human life across
the Arctic and subarctic possible as climates changed in the Late Pleistocene, allowing people to thrive in ecosystems that would otherwise
have been uninhabitable.[33] For these northern Indigenous groups,
caribou and reindeer were not only core to their material lives—but
also to their cultural, intellectual, relational, and spiritual lives.

Anthropologist Richard Nelson argues that close relations between
human and nonhuman have been vital to our evolutionary history:

> After all, for 99 percent of human history we lived exclusively as
> hunter-gatherers; by comparison, agriculture has existed only for a
> moment and urban societies scarcely more than a blink. From this per-
> spective, much of human experience over the past several million years
> lies beyond our grasp. Probably no society has been so deeply alienated
> as ours from the community of nature, has viewed the natural world
> from a greater distance of mind, has lapsed into a murkier comprehen-
> sion of its connections with the sustaining environment. . . . Imagine
> the whole array of North American animal species . . . each known in
> hundreds of different ways by tribal communities; the entire continent,
> sheathed in intricate webs of knowledge. Taken as a whole, this com-
> posed a vast intellectual legacy, born of intimacy with the natural world.
> Sadly, not more than a hint of it has ever been recorded. . . .

In Western society we rest comfortably on our own accepted truths about the nature of nature. We treat the environment as if it were numb to our presence and blind to our behavior. Yet despite our certainty on this matter, accounts of traditional people throughout the world reveal that most of humankind has concluded otherwise. Perhaps our scientific method really does follow the path to a single, absolute truth. But there may be wisdom in accepting other possibilities and opening ourselves to different views of the world. . . . A Koyukon elder, who took it upon himself to be my teacher, was fond of telling me: "Each animal knows way more than you do."[34]

From our nonhuman relations, Nelson wrote, we can learn "to recognize ourselves as animals rooted in the earth, beholden to the same rules of biology that govern all living things, equally dependent on the health and vitality of our environment, yet also burdened with special responsibilities because of our inordinate destructive power."[35]

In *Why Look at Animals*, John Berger argues that we have become human through our relations with other animals. The first animals arrived in human consciousness as messengers. We learned to be ourselves by observing other creatures and we became human by our relations not just with other people, but also with other animals.[36] Anthropologist Barbara King notes "[W]e think and we feel through being with animals."[37] Anthropologist Pat Shipman goes even further when she states that human evolution is not the story of one species figuring things out all on its own, or learning to control other species and force them into domestication, but rather "a collection of inter-species collaborations—between humans and dogs and horses, goats and cats and cows, and even microbes."[38] Many recent scholars have likewise argued that domestication was a reciprocal process, not an imposition of human will onto unwilling animal bodies. Other animals chose to enter into domestic relationships with people for the benefits this gave them, and both humans and nonhumans were profoundly altered by the process—on microbial levels, on genetic levels, and on cultural and spiritual levels.[39]

In contrast, most nineteenth-century European anthropologists assumed that the first step in human transformation from "barbarism to civilization" was learning to impose human will on other animals by forcing them into domestication. In this vision, hunting peoples were not as advanced as pastoral and agricultural peoples, and perhaps not even fully human. Anthropologist Hodder M. Westropp wrote in 1867 about "the reindeer race"—the Indigenous peoples who hunted wild reindeer. Reindeer people were "the primitive barbarous" who were the first stage in an "inevitable development from barbarism to civilization," Westropp argued. It was the duty of humans to "progress from a ruder stage to a higher and more advanced one," and "man" did that by learning "to domesticate his prey, and *reduce the wild animals around him to his rule*."[40] [Emphasis added.] J. W. Foster, then president of the American Association for the Advancement of Science, wrote in 1870 that Indigenous man was too busy hunting to "cultivate his intellect; and such culture, I need hardly affirm, is at the base of all civilization. How great the contrast between the primitive cave-dweller and the practical man of to-day. . . . The one was almost a brute; the other is almost a god!"[41]

For many nineteenth-century scientists, it made more sense to domesticate wilderness and "primitive, barbarian" tribes rather than to protect them. Federal programs were designed to save the Inuit from famine by substituting domesticated reindeer for the dwindling wild caribou herds. In their state of nature, Euro-Americans believed, Indigenous northern peoples would inevitably follow wild caribou into extinction unless they learned to herd domestic reindeer—or so the logic went, even though the reality was far more complex. Excellent scholars, such as John Sandlos, Lisa Piper, Bathsheba Demuth, and Roxanne Willis, have explored the dazzling complexities and failures of reindeer introductions into North America. Briefly, the projects didn't work. Even when the reindeer and their Sámi herders survived the difficult sea journey from Europe to North America, the introduced reindeer had little desire to stick around and often ran off to join the wild caribou herds that remained. And the Inuit had little desire to become pastoralists. Although several Sámi families prospered in Alaska and Canada, herding the reindeer that survived, little about the programs worked as intended.[42]

Similar logic was applied to proposals to save woodland caribou and woodland peoples from extinction, but the details differed in interesting ways. In the 1862 *Report of the Secretary of Agriculture*, Smithsonian scientist Spencer Baird noted that woodland caribou were diminishing across North America, even though they were still present in what he called "vast numbers" across Lake Superior's northern shore. When wild caribou populations declined across North America, Baird wrote, Indigenous people suffered terribly, leaving a "few half-starved, miserable Indians, in the depths of poverty and degradation." Rather than import European reindeer, he mused, why not domesticate woodland caribou instead? Domestication would place Indigenous woodland peoples "beyond the reach of those vicissitudes which are so rapidly sweeping off the Indians of the north and northeast of America." Bringing woodland caribou into a state of domestication would save the species by increasing private ownership and economic value, while learning to domesticate wild animals would also bring woodland tribes closer to civilization: "these Indians might become a pastoral people, and possibly, in time, as agricultural as the nature of the seasons would admit."[43]

Three decades later, as woodland caribou declines accelerated, other theorists suggested similar proposals. In his 1890 overview of *The Big Game of North America*, George Shields observed that "Doubtless, if turned to account, from his great strength, speed, and endurance, the Woodland Caribou of America could be domesticated, and his services made available in many ways advantageous to man."[44] In an interesting twist, Clarence Aldous, a wildlife biologist with the Lakes States Forest Experimental Station, released reindeer into the Superior National Forest on Minnesota's Lake Superior shore in 1930. (It's unclear whether these individuals were reindeer from Europe, or caribou from Canada.) The project quickly failed, because a "limited food supply" meant "many of the animals were lost."[45]

Diversity of Reindeer–Human Relations

The Euro-American fascination with domestication as a way to save both caribou and Indigenous peoples assumed that history went one way: from barbarism (hunting wildlife) to civilization (domesticating wildlife). Yet across the north, different human cultures developed a dizzying diversity of human relations with caribou: some domesticated, some semi-domesticated, and some fully wild. Rather than existing in a simple historical progression, from wild to tame, Indigenous peoples and caribou have been fluid in their choices, moving back and forth between wild hunting and domesticated herding, depending on a constellation of political and ecological factors that favored different strategies at different times.

In Siberia and Mongolia, Indigenous peoples domesticated some reindeer for food and transport, while hunting other herds of reindeer.[46] In Sápmi (northern Finland, Sweden, and Norway), the Sámi developed a semi-domesticated relationship with reindeer, shepherding them on their long migrations but never fully taming them as beasts of burden.[47]

North American caribou remain wild, and Indigenous peoples have neither domesticated nor tamed them. Yet the two species, human and caribou, have developed close material and spiritual relationships across all the north, regardless of the degree of domestication. For example, Gwich'in leader Sarah James said, "The Gwich'in are caribou people" who believe that "a bit of human heart is in every caribou, and that a bit of caribou is in every person."[48] All caribou peoples continue to recognize caribou as agents of their own, not mere beasts of burden, but sacred animals.[49]

Closer attention to a few of the granular details of caribou-human environmental histories suggests that there is no absolute boundary between wild and domesticated animals, and no absolute boundary between pastoral and hunting peoples. In Mongolia's taiga region, near the border with Siberia, the Indigenous Dukha people live nomadic lives with their reindeer—while ibex, sable, argali sheep, and other endangered wildlife roam the highest peaks. Conservation and tourism discourse often frames the Dukha as doomed, vanishing, the last of the true nomads—just as North American conservationists viewed both caribou and Indigenous peoples as doomed a century earlier. But are they really doomed?

The Duhka traditionally have lived in small family groups, migrating with their reindeer eight to twelve times a year, crossing what became the border between Mongolia and the Soviet Union to find the best forage for their animals. Rather than eating their reindeer, as the nomadic Sámi of Scandinavia and the Kola Peninsula do, the Duhka milk their reindeer and rely upon them for transportation. They include reindeer as members of a cohesive family rather than as mere domesticated animals (such as horses or dogs, which the Duhka also often have, but treat with less respect). Traditionally, they refused to eat reindeer (unless the animal was too old to survive by foraging). Instead they milked the cows several times a day and used the males as pack and riding animals.

The Duhka are shamanistic and reindeer play a core role in their spiritual lives. Rather than seeing reindeer as mere resources, the Duhka believe reindeer have agency and power—not just spiritual power, but also material agency in the taiga. For example, the Duhka invest one lead reindeer—the white stag—with mystical powers. That same stag chooses the routes to the best pastures and best forage.[50]

Similarly, reindeer nomads across Siberia created vibrant communities that flourished for millennia. Environmental historians Andy Bruno and Bathsheba Demuth and anthropologist Piers Vitebsky have described the effects of Soviet communism on Siberian and Kola Peninsula reindeer nomads, as the Soviets attempted to rationalize agricultural production, settle nomads, and outlaw shamanism.[51] These same political processes also affected the Duhka. When Stalin gained power, many Duhka fled to Mongolia, fearing they would be forced to give up nomadism and shamanism if they stayed in the Soviet Union.

In 2019, a colleague from the Smithsonian and I traveled to Mongolia to visit with several nomadic camps of Duhka reindeer people near the Russian border. After five days traveling by air, 4WD, and finally horseback, we reached the Duhka, or Tsaatan, as they are known in Mongolia, where each nomad family lives in a portable structure called an *ortz* (similar to a teepee), migrating with their reindeer across the taiga. Our goal was to witness the summer migration to high-elevation pastures, while learning how climate change, mining, and recent conservation policies have affected herding livelihoods. What we found from the Duhka surprised us. While they are concerned about climate change and mining, they believe misguided conservation policies have become a more serious threat to their reindeer and their culture.

Climate change in Mongolia is certainly posing grave risks to nomadic cultures. In the past seven decades, average Mongolian temperatures have risen 2.07 degrees Celsius, more than twice the global average increase of 0.85 degrees Celsius in the past century.[52] Warming has intensified summer droughts and extreme winter conditions. Over half the country's nomadic herders have been forced to abandon their herds in recent decades and move to the capital city of Ulaanbaatar, which lacks the infrastructure to provide clean water and clean heat to its more than 600,000 migrants. The result is burning of dirty coal, the worst winter air pollution in the world—and a spiral of intensifying climate migration, as more nomads lose their herds and migrate to the city.

Miners began exploiting the boreal forest for gold, jade, and uranium soon after the collapse of the Soviet Union in the 1990s. Poaching of snow leopards and other endangered wildlife increased, and the high alpine meadows, mountains, and streams that the Duhka see as sacred—and essential for their reindeer—were degraded by mineral exploration. The Duhka asked the Mongolian government to regulate miners, and in 2011 the government responded by setting up the Tengis-Shishged Special Protected Area, which cancelled all forty-four mining licenses in the region. But, concerned by poaching and habitat loss, the Mongolian government also eliminated hunting and made over 95 percent of the Special Protected Area off limits to reindeer—two changes that have sharply constrained Tsaatan livelihoods.

One of the hard lessons of conservation history is that national parks and protected areas across the world were often constructed at the expense of Indigenous peoples. Yosemite National Park, for example, was created by forcing out Indigenous tribes, casting them as enemies of wildlife.[53] Today the same process is happening in Mongolia—ironically funded in part by the Yosemite Parks Association, which has raised funds to pay the salaries of rangers.

Sustaining nomadic reindeer herders and protecting endangered wildlife requires that we re-think conservation policies that exclude human livelihoods from protected areas. Reindeer across the north have been a core part of human cultures for millennia, allowing humans to thrive after glaciers retreated. Today, reindeer actually help slow northern climate change, because their browsing decreases shrubs, thus increasing surface albedo, a measure of the reflection of the sun's heat.[54] Protecting reindeer and addressing climate change both require that we reconsider practices that exclude Indigenous peoples from full partnerships in designing conservation policies.

The Mongolian government is working hard to protect endangered wildlife such as snow leopards, but the Duhka insist their reindeer culture is equally endangered. If consulted as equals, they insist, they can help protect the boreal forest.

In Sápmi, the Sámi people continued to hunt rather than domesticate reindeer well into the sixteenth century. But when Europeans colonized Sápmi for mineral, forest, and agricultural resources, wild reindeer populations declined as both Sámi and Europeans began hunting them past their ability to reproduce. The Sámi developed a semi-domesticated relationship with reindeer to protect their remaining populations, shepherding them on their long migrations but never fully taming them as beasts of burden.[55]

A few days after Christmas 2012, I took the train south from Kiruna, a small mining town 200 kilometers above the Arctic Circle in Sweden. All three train cars were filled with Asian tourists drawn to Kiruna by the promises of shiny brochures: "See the Aurora Borealis in the last pristine wilderness in Europe! Come to Sweden's pure nature!" I stood at the window and watched the boreal forest slip by through the polar night. The moon rose over stunted spruce and birch trees bent beneath drifts of snow. From inside the warmth of the train, you could imagine the land outside as the frozen Arctic wilderness pictured in the tourist brochures, protected from human influence by bitter cold and distance.

But my husband and I had just spent the week with Sámi hosts on a farm in Puoltsa, a village 17 miles from Kiruna along the Kalix River. Even though daytime temperatures rarely rose above −33 degrees Fahrenheit, the landscape that looked so silent from the train buzzed with activity. Wood had to be chopped, guests had to be fetched, reindeer and horses had to be fed supplemental hay, snowmobiles had to be tinkered with. The taiga is anything but pristine wilderness: it is an inhabited landscape where the indigenous Sámi and their reindeer have lived ever since the ice retreated 10,000 years ago. Yet the wilderness perception persists, particularly now when a tourist boom makes it profitable.

Why does a flawed idea of wilderness matter? It renders invisible the Sámi. For centuries, urban governments have used the idea of the boreal north as uninhabited and remote to promote colonization of the north for its resources. Open-pit iron mines proposed for Sámi territory in the ore-rich landscape near Kiruna, Sweden, continue to be justified by similar logic. The Kiruna region contains the largest underground iron-ore mine in the world; 90 percent of all iron ore mined in Europe comes from here. From the Swedish government's perspective, mining is inevitable because the world needs iron ore for steel, and the government needs mining profits to fund the social programs integral to Swedish society. But from the Sámi perspective, the proposed mines would make it impossible to continue reindeer herding, ending thousands of years of successful cultural adaptation to the taiga (see plate 4).

Sámi herders do not just follow the reindeer. They have negotiated complex relationships with the animals, basing an eight-season migration pattern on the reindeer's seasonal cycles. From March until April, the pregnant female reindeer begin to move out of lowland forests toward mountain pastures. In April and May, calves are born, and the cows choose spring pastures where snow melts early, allowing them to supplement their lichen diet with leaves, grass, and herbs. In June, reindeer select riparian vegetation near marshes and brooks, allowing them to quickly regain weight lost during the long winter. From June to July, when parasitic flies become a problem in the lowlands, reindeer move up into high, windy mountain meadows where they can find relief from both insects and heat. When August comes, reindeer build up fat and muscle to prepare for the winter and begin their migration to lower pastures, where the rutting period commences. During the rut, cows continue to accumulate reserves for the winter, while the bulls expend much of their stored body fat and muscle weight, which they will need to replenish during the brief autumn season. When the snows deepen in early winter, the reindeer migrate back toward lowland forests where they can dig through the snow cover to reach their main winter food source: ground lichens. From December through March, the coldest season, reindeer graze for lichens and berry plants in the coniferous forests. These migration patterns paid little attention to national borders; to find the best forage, reindeer often crossed from summer pastures in Norway into winter territories in Sweden and Finland.

Mines now proposed across Sápmi would alter Sámi relations with reindeer within their territories, making it impossible, the Sámi fear, for reindeer herding to persist. The mines, for example, would likely consume 60 to 70 percent of the spring pasture area and substantial portions of fall pasture. Mining infrastructure would block critical migration routes, and dust from tailing piles would change succession patterns in spring pastures, allowing grass to overtake ground lichen. Autumn pastures would also be reduced, increasing reindeer vulnerability to harsh winter conditions. From the Sámi perspective, how can 20 years of mining take priority over thousands of years of Sámi culture?

The Swedish government contends that the Sámi have no rights to exclude competing uses from Sámi territories. From their perspective, industrial development is inevitable and the Sámi must make way for it. If the good ore happens to interrupt a migration route, then move the reindeer somewhere else. Put them in trucks if necessary. If tailings piles eliminate lichen, then feed the reindeer something else.

In this view, domesticated reindeer are essentially cogs in a machine, not members of interconnected ecological systems. The logic assumes that sharp boundaries exist between wild and tamed nature. But Sámi relations with reindeer disrupt these boundaries, for the reindeer remain individuals with agency, not mere machines. The migration paths reindeer choose are negotiations with the Sámi herders, not engineered decisions imposed by technical logic. The food they eat is neither purely wild nor domestic, and the pastures they need cannot simply be replaced.

Lessons from Caribou

What can we learn from caribou-human relationships across the globe to help us forge resilient relationships for the future? One lesson is clear: we live in a world of relationships. Humans, like other animals, are embedded at every point in the environment. For millennia, caribou and people have wandered the earth together, on meandering migrations. Sustaining those connections is important to sustaining not just woodland caribou, but also people.

Americans typically imagine caribou as creatures of distant wilderness, a remnant of primeval nature that was irrevocably lost to industrialization. Woodland caribou, in this discourse, need vast, untouched wilderness and will be doomed by climate change and the Anthropocene. These beliefs are powerful, but they are flawed—and they let government agencies avoid taking pragmatic actions today to restore woodland caribou. There's nothing inevitable or mysterious about the demise of woodland caribou. Specific policy decisions led to their declines in the twentieth century, and reversing those policy decisions has the potential to lead to their rebound, but not if we continue imagining woodland caribou as creatures of an untouched primeval forest.

Current rhetoric about woodland caribou mirrors the rhetoric of early conservationists in the late nineteenth century. When agencies and NGOs talk about woodland caribou as too vulnerable to be sustained in the Anthropocene, it becomes a self-fulfilling prophecy. Agencies and environmental groups become reluctant to restore them, for that suggests continuing care, a need to keep investing time, resources, and energy to manage predators and migration routes. For woodland caribou to thrive, we need to rethink assumptions about the need for continuing human stewardship of migratory wildlife in a warming world. People and caribou have long had intertwined histories, and caribou may now require continuing care, a need to keep investing time, resources, and energy to manage their predators and migration routes.

As mainland caribou populations vanish, and the species becomes a ghost in the memory of residents, the conviction that the caribou are a core part of Indigenous culture has been vanishing as well among some of the Indigenous bands along the northern shore of Lake Superior. They've been reluctant to do what it might take to sustain caribou populations, if that means decreasing energy development or intensive forestry, or culling moose and reducing their revenues from moose hunting. Other bands, such as the Michipicoten First Nation, are pushing hard for caribou restoration, but the intertribal dynamics become complicated, with different bands gaining revenues from extractive projects that other tribes do not participate in.

As the Pukaskwa National Park superintendent made clear when she refused to entertain a proposal to translocate caribou to the park, the government's goal is for a self-sustaining population that would not need continued human intervention. Caribou advocates, including some of the First Nations leadership and bands, have a different goal: it is fine to them to continue investing in caribou translocations, if that means the population averts extinction. The caribou translocated to Michipicoten, the Slate Islands, and the Gargantua Peninsula may not have persisted forever, but they did persist for decades. This suggests that continued efforts may be required. A helicopter ride here, a wolf cull there: if that's what it takes to sustain a species through the coming disruptions, who decides whether that's worthwhile?

Biologists Lauren Anderson and Zeke Hausfather write in The Breakthrough Institute's magazine: "As climate change begins to shift habitats and change the range that species occupy, protected lands and wildlife corridors—though critical tools—won't always be sufficient. . . . If species are not able to naturally disperse as their ranges shift in a changing climate, wildlife managers may need to take a more active role in physically introducing them into new areas.[56] This violates a particular ideal of wilderness beloved by many conservationists. Indeed, the goal of creating self-sustaining populations that won't need continued human care is reflected in Canadian wildlife policy. Because certain quantitative models suggest that woodland caribou will not be able to create a self-sustaining population on the Lake Superior mainland in the near future, the Pukaskwa National Park staff believe it is unethical to move forward with translocations. But the caribou advocates find the park position equally unethical, because they believe that the Euro-American concept of nature separate from people motivates decisions that will lead to extinction. Rather than assume Euro-American ethics are necessarily correct, the caribou advocates urge ministry staff to involve Indigenous communities as full partners in the restoration of caribou.

Sustaining caribou and Indigenous cultures across the north requires real partnerships, across cultures and across species. Protecting caribou and addressing climate change both require that we reconsider practices that exclude Indigenous peoples from full partnerships in designing conservation policies.

Caribou will indeed dwindle in a warming world if we restrict their migrations and refuse to manage their predators. But climate change should not be an excuse to give up on the management strategies, here and now, that could keep them from extinction. Climate change isn't going to doom woodland caribou. Human policy decisions, however, might.

FOUR

Indigenous Communities and Lake Sturgeon Restoration

ach day I drive over the Sturgeon River on my way to work. On class field trips, we canoe through the Sturgeon River Slough, winding our way through backwaters where great rafts of ducks land to feed and rest during their fall migration. In the winter, my husband and I snowshoe into the Sturgeon River Gorge Wilderness to admire the frozen Sturgeon Falls. Next summer, if social distancing finally ends, I hope to check out the restored wetlands in the Sturgeon River Sloughs Wildlife Area with a wetland ecologist.

These place names tell us that the lake sturgeon (*Acipenser fulvescens*) was once an important fish here. As sturgeon made their annual migrations, both Indigenous cultures and European immigrants came to depend upon their flesh for material sustenance. For Indigenous communities, sturgeon meant much more than simply food: their migrations helped to weave the world together, spiritually, culturally, and relationally. But few of us are lucky enough to bump into these sea monsters now when we swim and fish and paddle in the waters that bear their name. This chapter explores the collapse and (partial) recovery of lake sturgeon in the upper Great Lakes. Why did they decline, and which restoration efforts are helping their recovery?

FIGURE 4.1 Settlers targeted lake sturgeon for eradication in the nineteenth century because the sturgeon had little commercial value and damaged whitefish nets. (Photo by J. Kitchin, 1900.) *Minnesota Historical Society*

Sturgeon are astonishingly big fish (see plate 5). In the Great Lakes, lake sturgeon can reach over 7 feet long and 330 pounds in weight. Its cousins in the Black Sea can reach 4,400 pounds, more massive than a beluga whale or a narwhal.[1] Their life span was once far longer than that of humans. Some writers call these huge fish "leviathans," after the sea monster in the Book of Job that terrified generations of Bible readers. As angler and writer John Waldman notes, they're "a whiskered relic of a group that dominated the seas during dinosaur times, but is now reduced to two-dozen species worldwide, with many in peril."[2] In much of their former range, lake sturgeon are now ghosts.

There's hope in this particular ghost story, however. While lake sturgeon have a long way to go for full recovery, they've been pulled back from the brink of extinction by a diverse collection of Indigenous groups, fisheries biologists, commercial fishing companies, and recreational anglers.

Wisconsin started conservation efforts to stem the decline of sturgeon in 1912, and other Great Lakes states and Canadian provinces soon followed suit. Lake sturgeon are now federally protected in the United States, listed as either threatened or endangered in nineteen of the twenty states within their original range, and they also have protections in Canada. Yet for a century these conservation efforts failed to stem their decline. Things changed after the tribes in the Great Lakes region won federal recognition of their reserved treaty rights to comanage fisheries on ceded territories. Once that happened, the tribes could lead restoration of the fish the Anishinaabeg know as *nmé*, bringing back their ancient clan partners to the waters of the north. That story of Indigenous-led restoration offers hope as diverse communities try to avert extinction of migratory species in the upper Great Lakes.[3]

Sturgeon Life Histories

Lake sturgeon may live for more than 120 years, which means their life spans can straddle that of three generations of people. But their longevity makes them vulnerable to overharvest, because they don't mature and begin to breed until they are decades old. Killing an adult female slices off a substantial chunk of breeding experience, severing the future of the species from her past.[4]

Biologists sometimes call sturgeon "living fossils," for they have remained largely unchanged for over 150 million years, when dinosaurs still roamed the earth. Historian Richard Carey writes: "Until just the last fraction of a second in geologic time, sturgeons were the dominant large fish in every major river system on all three continents of the temperate northern hemisphere. . . . The sturgeon is geologic time made flesh, and the length of its tenure on earth is impossible to comprehend."[5]

Sturgeon as a group are among the oldest life forms still living on earth. Their lineage began to evolve soon after the devastations of the end-Permian mass extinctions nearly 250 million years ago. Their ancestors somehow took advantage of this bleak post-Permian world, and thus began their evolutionary journey as the oldest of the bony fishes. About 150 million years ago, their current forms appeared, and since then they have changed little, lending them the name "living fossils."[6]

Lake sturgeon have a fragmented distribution in North America. Before Euro-American industrialization, sturgeon were found in the Mississippi River drainage basin south to Alabama and Mississippi. They also swam in the Great Lakes and the Detroit River, east down the St. Lawrence River to the zone where freshwater meets the ocean. Lake sturgeon range as far west as Lake Winnipeg and the North Saskatchewan and South Saskatchewan Rivers. In the north, they once made it clear to the Hudson Bay Lowlands, at the edge of the

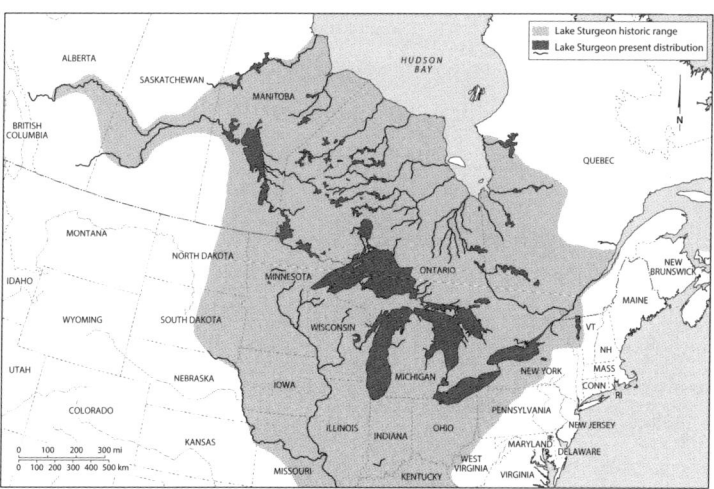

MAP 4.1 The lake sturgeon's historic and current range in North America. *Bill Nelson*

Arctic. In the east, lake sturgeon swam in Lake Champlain and in some Vermont rivers.[7]

In each of the major watersheds where they remain, they are what biologists call "endemic"—meaning that their populations are essentially isolated from each other. Sturgeon in the Great Lakes migrate 300 miles or more on their annual circuits of the near-shore, but they never interact with sturgeon in the Hudson Bay watershed or the Mississippi River watershed. (This endemism has implications for modern restoration: Euro-American biologists learned from tribal elders to only use brood stock coming directly from the same watershed.)

It's difficult to make sense of this fragmented distribution without understanding some of the glacial history of the continent. As the great ice sheet crushing the continent melted during warmer interglacial periods, a tremendous lake of meltwater formed, held back by chunks of rock and ice. In its earliest stages, Glacial Lake Agassiz was most likely a "cold sterile lake with unvegetated margins." But a few marine animals survived the glaciation, and they somehow found their way into the lake. Remarkably, several species of small mollusks and "the bony plate of a sturgeon have been found in the beach deposits of Lake Agassiz."[8]

Each time the ice dams broke, a rush of cold freshwater poured across the continent, carving out much of our modern Great Lakes geology, crashing through what's now the St. Lawrence Seaway, Hudson Bay, even the Mississippi River. The great floods distributed lake sturgeon into these now-separate watersheds across the continent, where they evolved in isolation over thousands of years.

Sturgeon and caribou faced similar challenges during the Pleistocene, a period of rapid climate change. When boreal watersheds became glaciated, populations adapted by moving south and east to Mississippian and Atlantic refugia. During the warming periods that followed the end of each glaciation, species returned northwards.

In the face of these geological dramas, sturgeon evolved what ecologists call "resilience." Resilience is the capacity of an ecosystem or species or community to recover from disturbance, and persist—changed, but still recognizable, maintaining its normal patterns of nutrient cycling and biomass production. For sturgeon, resilience meant that they were able to adapt to dramatic temperature fluctuations, floods, and changing food sources, with great food abundance at some times and little food at other times. One of their core strategies for resilience was migration. When spawning time came, adult sturgeon could migrate hundreds of miles into tiny streams, where their eggs and fry (the term for baby fish) could find shelter from ravenous predators. Later in the season, instead of starving in the small streams that offered protection at spawning time but little food, the sturgeon could migrate to open lakes with abundant food supplies.

Sturgeon are enormous creatures, and they need a lot of food to sustain their bodies. They also lack teeth, which means they can't chomp down on little fish the way many big fish do. Instead they have a tube-like mouth that protrudes from their nose (see plate 6).

They suck their food up from the lake bottom, like giant vacuums "grazing on aquatic invertebrates and mollusks."[9] They can swallow a dead salmon whole. As vacuum feeders, they have figured out a way to "advance into an ecological niche unexploited by other fish."[10]

Little streams don't have a lot of resources for sustaining really big fish, so sturgeon head to enormous lakes, swimming at depths of 15 to 30 feet near the shore, where nutrients are concentrated. They feed along lake bottoms, consuming vast quantities of snails, crustaceans, aquatic insects, mussels, and small fish. These strategies allow them to concentrate big stores of energy in their slowly growing bodies.

While feeding, lake sturgeon are also famous for a curious habit called vaulting. They leap completely out of the water and come down with a massive belly flop, thwacking the water so loudly that you can hear it from "more than a half-mile away and probably farther under water." In 1731 an observer marveled that in the spring and early summer, "the rivers abound with [sturgeon], at which time it is surprising, though very common to see such large fish elated in the air, by their leaping some yards out of the water; this they do in an erect posture, and fall on their sides, which repeated percussions are loudly heard some miles distance."[11]

This is surprising behavior because it takes a lot of energy to vault from the water, especially when you weigh hundreds of pounds. Evolutionary biologists argue that costly behaviors should vanish from a population unless they have an evolutionary benefit. What might those benefits be for a leaping lake sturgeon? Biologists have suggested a litany of possibilities, including group communication, courtship display, predator avoidance, or removal of parasites. Or we can anthropomorphize this behavior, speculating that, as one fisheries biologist suggests, perhaps it "simply feels good."[12] Whatever the answer turns out to be, the very existence of behavior that seems so puzzling to Euro-Americans suggests how much biologists still have to learn about creatures who sense the world in ways so foreign to ourselves.

When lake sturgeon approach sexual maturity (typically after 24 to 26 years for females; half that for males), they begin to make their migratory journeys, swimming upstream hundreds of miles to their natal (birth) streams. A female who lives long enough to breed must navigate past the dams, fishing nets, and pollution that bar her path. If she makes it through that obstacle course and reaches her birth stream, she can lay up to 3 million eggs in one season. Unlike some species of salmon, who dig a careful nest or "redd" in the gravel and then die after spawning, a female sturgeon doesn't bother with a nest. Instead, she broadcasts her eggs over shallow rocks or gravel then swims back downstream to the open lakes. If she remains healthy, she may return again to spawn in future years.

What happens to those millions of eggs at the bottom of the stream? Most die, but some are fertilized by male sturgeon. Some of those fertilized eggs will survive long enough to develop into tiny larval fish, too small and weak to feed or swim on their own. They survive off the nutrients in their yolk sacs, drifting downstream upon the currents until they eddy into protected backwaters—the oxbows and sloughs that are the legacies of a river's history upon the landscape, the river's palimpsest.

Oxbows provide a sort of nursery where the tiny sturgeon develop into fry that can start feeding on insect larvae. A few of those fry few grow big enough (7 to 8 inches long) to swim back into the main river and determine their own direction rather than relying entirely on the currents. They head for open waters, where they can slowly grow to vast sizes on the abundance of the big lakes.[13]

Human-Fish Relations: Meanings of Sturgeon for Indigenous Peoples

Native American tribes from across northwest Minnesota, the Dakotas, and portions of Canada have long honored the lake sturgeon as an essential staple and a sacred relative (see plate 7). Fish migrations

shaped human cultural patterns, clan governance, and seasonal movements in the upper Great Lakes. For the Rainy River First Nations,
the Red Lake Band of Chippewa Indians (described in the caribou
chapters), and the White Earth Band of Ojibwe, the spring sturgeon
harvest was a time to gather after a long, isolated winter and celebrate
the return of spring—and the return of the sturgeon. As White Earth
Band anthropologist Tom McCauley told scholars Marty Holtgren,
Stephanie Ogren, and Kyle Whyte, the spring sturgeon migrations
were profoundly important: "So ancient is this fish species, and so
deeply intertwined is it with the tribe's culture and survival, that the
Anishinaabek sometimes call it the 'grandfather fish.' But usually tribal
members refer to the fish, which has been around since dinosaurs
roamed the earth, as *nmé*."[14]

Lake sturgeon were once a core part of tribal identities across the
upper Great Lakes and north to Hudson Bay. The Gun Lake Tribe
of Potawatomi speak of rivers once so full of sturgeon that a "person
could walk across water on the backs of the fish without wetting their
feet." The Ojibwe in White Earth, Minnesota, have similar stories,
telling of "the fish being so plentiful that it appeared to them you
could walk across the river on the backs of the fish."[15]

For the Anishinaabeg, sturgeon migrations figured prominently
in their human migrations from the St. Lawrence Seaway. When
Anishinaabeg peoples migrated westward from the mouth of the St.
Lawrence River into the Lake Superior basin, they were not only
fleeing violence in the Eastern Seaboard, but were also drawn by
visions of *manoomin*, the "food that grows on water," a type of wild
rice (*Zizania palustris*) that was once abundant in the region.[16] Lake
sturgeon were the keepers of the wild rice, which made them all the
more important to the first peoples. The Anishinaabeg brought the
clan connections—and smoked flesh—of lake sturgeon with them to
sustain them on their journey. When they reached the upper Great
Lakes, they gathered on the banks of sturgeon-bearing rivers to cel-

ebrate the annual return of sturgeon to their spawning grounds, and to celebrate their own safe arrival.[17]

When the French fur adventurer Pierre Radisson arrived in Chequamegon Bay in 1654, he mentioned how abundant and important the sturgeon were for the northern tribes. "There is a channel where we take a great store of fish: sturgeons of vast bigness and pike seven feet long. At the end of this bay we land. The wildmen give thanks to that which they worship, we to the God of gods," wrote Radisson.[18]

A little more than a decade later, Father Claude Allouez described the religious beliefs of the Anishinaabeg along Chequamegon Bay. In Anishinaabeg beliefs, sturgeon were sacred because they connected the living and the dead, the human and the nonhuman, all relations worthy of respect. But Allouez found this horrifying because it violated his Christian sense of the proper boundaries between human and nonhuman. He complained: "They believe the souls of the dead rule the fishes of the lake. From earliest times they have believed in the immortality and reincarnation of the souls of dead fishes. They never throw fish bones into the fire, for fear they might offend the souls, and the fish will no longer enter their nets. They hold in great awe a certain fabulous animal which they never see except in dreams, and which they call *Missibizi*. They recognize it as a great spirit, and offer it sacrifices to obtain good sturgeon fishing."[19] In another Anishinaabeg tribe, the Gun Lake Potawatomi, "Sturgeon are revered as grandfathers and grandmothers."[20]

For the White Earth Ojibwe, "The members of the Fish Clan, and subsequent Sturgeon Clan, are the presumed descendants of the first beings to rise from the water. The sturgeon is considered to be the spiritual keeper of the fisheries. At the Lake of Woods, Minnesota, spawning ground, missionaries were discouraged from developing missions at the spawning grounds as it was believed that this could destroy the sturgeon fishery."[21] The sturgeon, in other words, were creatures with agency and spiritual power who paid close attention to their human relatives, and who could react with anger if people violated their spiritual agreements.

From 1668 to 1670, Bacqueville de la Potherie reported on the Sturgeon Clan among the northern tribes, who had gathered to meet with the explorer Perrot: "At the end of May, the chiefs of Green Bay, those of Lake Huron and Lake Superior, as well as the people of the north and several other tribes all came to the Sault . . . [Perrot] drew up an official report, and he made all the peoples sign it. The chiefs, for their signatures, drew the insignia of their families. Some drew a beaver, others an otter, a sturgeon, a deer or an elk."[22]

The clans mentioned here are lines of lineage in Anishinaabeg culture. Clans are named for native birds, animals, and fish, and they honor specific characteristics of each species. Nmé, or Sturgeon Clan, is the *ogema* or chief clan of the fish pantheon. Fish Clan people are "considered the spiritual advisors and mediators of the Tribe; they are considered the philosophers and spiritualists. Like the fish themselves, Sturgeon Clan members have the longest lifespan amongst all the clans."[23] Clans are core to governance for the Anishinaabeg, with each clan serving a unique purpose for the entire community. The Sturgeon Clan helps to make important community decisions. For example, before the spawning runs of the sturgeon begin in the spring, "tribal leaders consult with the sturgeon clan on fishing locations to avoid over fishing."[24]

Nmé populations in different parts of the watershed are associated with different clan identities. For example, different places along each river are set for different clans to fish. "Just as Anishinaabeg families today are descendants of generations of Anishinaabeg from this region, so too are surviving nmé today the descendants of those who interacted with the very same families generations ago. The fish was so important to the traditional culture that Anishinaabeg leaders would sign documents with nmé images."[25] For the Anishinaabeg, sturgeon are both clan leaders and key food sources. The Ottawa speak of "The grandfather fish [who] would sacrifice itself so the people would have food [during the lean seasons] until the other crops were available."[26]

In the late eighteenth century, when John Tanner lived among his Anishinaabeg family in the north during a time of great cultural change, sturgeon continued to play central roles: as food, as spiritual connection, and in clan governance. As noted in chapter 2, Tanner wrote that, on one occasion, his family decided to paddle from Grand Portage across to Isle Royale (a 15-mile crossing in treacherous waters) "where, [a man] said, were plenty of Caribou and Sturgeon, and where, he had no doubt, he could provide all that would be necessary for our support. . . . We also took, with spears, two or three sturgeons immediately on our arrival; so that our want of food was supplied."[27]

When Tanner killed his first sturgeon, Net-no-kwa—the matriarch of Tanner's adopted family (who Tanner called the "old woman")—demanded that he mark the event with a spiritual ceremony. "[A]s this was the first sturgeon I had ever taken, the old woman thought it necessary to celebrate the feat of Oskenetahgawin, or first fruits, though, as we were quite alone, we had no guests to assist us."[28] Tanner was nervous about stopping for the ceremony because "The mouth of the Assinneboin is a place much frequented by the Sioux war-parties, where they lie concealed and fire upon such as are passing." But Net-no-kwa insisted that the ceremonies were more important than hiding from the Sioux, and indeed they did run into some difficulties that evening. When Tanner traveled to the Rainy Lake region in Minnesota (near the current border between Canada and the United States), he found that the falls upon the river, where sturgeon gathered on their spawning journeys, were places where he could catch enormous old sturgeon. He wrote: "The river which falls into Rainy Lake, is called Kocheche-se-bee, (Source River,) and in it is a considerable fall, not far distant from the lake. . . . One day, as I was fishing here, a very large sturgeon come down the fall, and happening to get into shallow water, was unable to make his escape. I killed him with a stone, and as it was the first that had been killed here, Sah-muk made a feast on the occasion."[29] Another time, at the rapids of Rainy Lake River, he took 150 sturgeon to help prepare for the lean months of winter.[30]

Sturgeon were of such importance to the Anishinaabeg that, when Tanner was in despair over the death of his child from measles and the hunger of his family, he dreamed that sturgeon came to save him from hunger and sadness. Caribou and waterfowl came as well to greet him in his dreams and rescue him from suffering. He wrote: "I looked and saw before me many ducks covering the surface of the water, and in another place a sturgeon, in a third a reindeer. This dream was fulfilled, as usual, at least as much of it as related to my hunting and fishing."[31]

European Invasions, Accelerating Losses

The Anishinaabeg for generations gave dried sturgeon as gifts, rather than using it as a trade item. Sturgeon gifts helped weave threads of relationship between different families and bands. But as Europeans invaded tribal territories, the tribes used sturgeon to negotiate better trade relations and then treaties. Sturgeon "had become an important trade item when working with Europeans. This was especially shown when the primary food for Europeans was traded sturgeon meat and the tribes could use this to their advantage in trade agreements." Europeans moving into Anishinaabeg lands, such as Radisson and later immigrants, needed food, not just beaver or the promise of gold. Dried sturgeon became the most important source of protein they could get from the tribes. Michael Kallok of *Minnesota Conservation Volunteer* notes that "because of this abundance, early Europeans regarded these Ojibwe as difficult trading partners."[32]

Settler colonialism changed these relations. The largest sturgeon fishery along Minnesota's border with Canada had long been controlled by the Anishinaabeg.[33] While European traders had been willing to trade for dried sturgeon on terms favorable to the tribes, Euro-American settlers demanded exclusive access to the sturgeon spawning sites, squeezing tribes out. Starting in the second half of the nineteenth century, after commercial fishing groups and settlers pressured governments, Canadian and American jurisdictions passed numerous laws limiting the treaty fishing rights of Indigenous communities.

Because European immigrants had initially relied on lake sturgeon for food, they saw some value in them as resources, if not as relations. But as an economy based on commercial whitefish and lake trout fishing came to replace the declining beaver trade, whites began to perceive sturgeon not as a valued resource, but as a pest to be exterminated. Sturgeon ripped whitefish nets, so commercial fishermen simply slaughtered them, tossing the bodies on shore to rot, leaving great stinking piles that horrified tribal members.

U.S. Fish Commissioner James Milner wrote in frustration about the terrible waste of sturgeon, "destroyed in the most wanton and useless manner," because they damaged the whitefish nets.[34] In Green Bay, Wisconsin, he said, whitefish nets set during the fall often contained "a hundred or more" sturgeon that "are considered a nuisance and annoyance. A few fishermen are considerate enough to lower the corner of a net and allow them to escape, but the commoner way is to draw them out of the net with a gaff-hook and let them go wounded, or to take them ashore and throw them on the refuse-heap, asserting that there will be so many less to trouble them in future. A very large number are destroyed in this way."[35]

In the latter half of the nineteenth century, however, new markets emerged for lake sturgeon.[36] A growing steamship trade started running out of riparian trees for fuel, so the captains turned to the oily flesh of sturgeon. Rather than tossing carcasses back into the water, commercial fishermen piled dead sturgeon up like logs along the river, to stoke the furnaces of steamships.[37] Toss a few of the huge fish into the belching stoves, and you could chug your way—stinkily—upriver.[38]

Because a growing beverage trade in America needed isinglass for clarifying beer and wine, sturgeon also became sought after for their swim bladders, a traditional source of the substance. For relatively small quantities of isinglass, however, enormous fish were left to rot.[39] But even more detrimental to North American sturgeon populations were the new international trade networks in caviar. When the European caviar market experienced steep increases in the prices for beluga sturgeon roe, a few enterprising traders turned to North America, caught lake sturgeon, sliced their ovaries open, and shipped their eggs to Europe labeled as "Russian caviar." Milner wrote that "the perception of lake sturgeon among commercial fishermen" changed once a caviar plant in Sandusky, Ohio, had created a market for the fish. Kallok calculates that "Lake sturgeon harvest topped 1 million pounds a year at the south end of Lake of the Woods during the late 1800s."[40] At prime sturgeon spawning sites on the Great Lakes, such as Grassy Meadows in western Lake Erie, Euro-American fisheries and villages popped up, forcing out tribal fisheries that had used those same sites for many generations.[41]

By 1873, sturgeon populations were noticeably declining. Milner wrote that "The decrease in numbers is apparent, to a certain extent, in localities where the pound-net has been in use for a number of years. At Sandusky, Ohio, the numbers brought in from the nets and handled at the curing-establishment in a season are said to have nearly reached eighteen thousand a few years ago, while in 1872 the books showed a record of thirteen thousand eight hundred and eighty received."[42] Milner went on to vent his frustration at the ghost nets that were killing fish long after careless commercial anglers abandoned them. He wrote "of great numbers of fish destroyed uselessly by the breaking away, in storms and currents, and loss of nets in the lakes, which continue to capture fish until the floats become water-logged and sink to the bottom."[43]

The loss of fishing rights—rights that had been guaranteed in treaties but ignored over a century of Bureau of Indian Affairs gover-

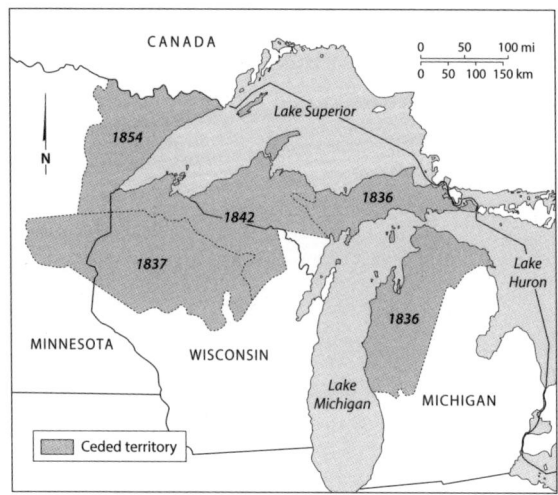

MAP 4.2 Treaties signed with the Anishinaabeg and
ceded territories in the Lake Superior basin. *Bill Nelson*

nance—accelerated the sturgeon's collapse. Newer treaties in the nineteenth century further opened the upper Great Lakes for mining and
logging, placing growing stresses on free-flowing rivers and streams.

Deforestation in the late nineteenth and early twentieth centuries
proved particularly damaging. As I have noted in *Sustaining Lake
Superior*, between 1870 and 1909 the lumber industry boomed in Wisconsin, Michigan, and Minnesota, providing a third of the nation's
forest products. By 1898, only 13 percent of the white pine in northern Wisconsin remained. Harvest practices and slash fires increased
erosion and siltation in spawning habitat, while splash dams and log
drives damaged riparian areas. Stream temperatures rose as forest
cover was removed, reducing spawning habitat as well. Tremendous
quantities of sawdust were dumped in rivers and estuaries, blocking
fish passage. Many contemporary observers noted how, as forests
fell to the axe, lakes and streams became muddied, water dwindled
in streams and wells, and floods became more common, along with

summer drought. These combined effects likely reduced spawning and rearing habitat for migrating sturgeon.

In the first decades of the twentieth century, a new paper industry boomed on the Canadian shore of Lake Superior. Companies dumped their waste products into waterways, contaminating spawning sites, depriving migrating fish of oxygen, and generally stinking up the place.[44] In Minnesota, effluent from paper mills obliterated prime sturgeon spawning areas in the Rainy River. "Were it not for the Clean Water Act of 1972," Kallok argues, "lake sturgeon would have likely disappeared from this border water between the United States and Canada."[45]

Even more insidious than paper pollution were the hydropower dams created to supply industry needs. Dams concentrated migrating sturgeon, typically the oldest individuals, making them much easier to harvest. Pollution also concentrated below the dams, creating a lethal cocktail for the fish. Those individual adults who managed to escape the harvest and the pollution kept trying to migrate to their birth streams, but most failed. Some died, while others swam back to the big lakes, unable to reproduce.

By the 1920s, the proliferation of hydropower dams had created complicated international issues in the Great Lakes basin. As the Canadian pulp industry boomed, thanks to federal and provincial investments to build infrastructure and relax water-quality requirements, an over-production of paper was the result. This led to an economic crisis in the paper industry, with prices tumbling. Rather than reduce paper production, the Canadian government decided to increase hydropower capacity even more, hoping to outcompete American paper, thus driving American paper companies into bankruptcy. Then, monopoly in hand, the Canadians could force prices up again to pay back their loans and recover their fortunes.[46]

Engineers developed a plan to accomplish this—a plan that was breathtaking in its own way. The Great Divide—a height of land that lies just north of Lake Superior and Lake Nipigon—is where freshwater flows either south into Lake Superior and the Great Lakes, or else north to the salty Hudson Bay. The engineers asked: Why not divert water that would otherwise flow into Hudson Bay and send it south into Lake Superior to power increasingly massive dams? As one columnist wrote in a November 1925 issue of the *Chicago Tribune*: "Dam it all, the engineers say, referring to the rivers that braid north of the divide and empty into James Bay, below the Hudson. The hydro-electricity produced and the improved navigation on deep water in the Great Lakes should quickly offset the immense costs, the men insist. . . . Every cubic foot of water diverted from the relatively useless basin of Hudson Bay into the lakes will prove liquid gold."[47] Chicago engineer H.P. Ramey proposed creating a vast reservoir by damming the river system and raising water levels up to the height of land. Then he would dig a tunnel through the divide and channel water once destined for Hudson Bay into Lake Superior. This vast engineering scheme would solve "the Great Lakes problem," planners promised.[48]

The 1925 proposal disintegrated in the tangle of competing federal and state jurisdictions. But even without that grand re-engineering of the basin, sturgeon harvests collapsed. Between 1879 and 1900 the Great Lakes commercial sturgeon-fishing fleet caught an average of 4 million pounds of fish per year. In one single abundant year, 1885, 8.6 million pounds of sturgeon were harvested from the Great Lakes. By 1928 the catch totaled only 2,000 pounds. "The fish that had abounded in drainages of Hudson Bay, the Great Lakes and the Mississippi River as far south as Alabama suddenly faced extinction."[49] By the early 1920s sturgeon were extirpated from the Red River of the North as well as the St. Louis River and the western basin of Lake Superior. The last known native-born lake sturgeon from the upper Minnesota River watershed washed up on the shores of Big Stone Lake in 1938.

As these losses accelerated, a host of conservation efforts accelerated

FIGURE 4.2 Chicago engineer H. P. Ramey's 1925 plan to change the Great Lakes (from the *Winnipeg Evening Tribune*, November 6, 1925).

as well. In 1912, Wisconsin banned commercial fishing of sturgeon in the Menominee River. In Minnesota, a 1920 law required sturgeon less than 15 pounds to be released. In 1947, the Minnesota Department of Conservation enacted a daily limit of one sturgeon 30 inches or longer.[50] But these efforts didn't deal with core problems: dams, pollution, destruction of spawning habitat, and the loss of cultural connections with Indigenous protectors.

Fisheries managers were horrified by the repeated collapse of one fishery after another. They wrote endless studies asking the same basic question: How to rescue collapsing fisheries, when you couldn't challenge the relentless calculus of economic models that valued development over tribal rights and healthy watersheds? The answer seemed obvious: use aquaculture to avoid what they saw as the inherent waste in natural systems. If a female sturgeon spawned millions of eggs in a single go, but only one or two survived to adulthood, surely that was an enormous waste? Borrowing techniques from western salmon fisheries, they focused on hatcheries and stocking.[51] Yet hatchery programs in isolation failed to address the multiple stressors that were destroying migratory fish. As with salmon hatcheries in the western United States, scientific beliefs about the ability to control the biological productivity of fish and improve upon natural processes influenced the decision to focus on stocking as a management tool rather than harvest limits or habitat restoration.

Cultural Losses

For settlers, sturgeon's absence cost a few fishermen their pay, but the benefits attached to hydropower far outweighed those losses. The rational cost-benefit analyses that lined the shelves of government agencies all shared an implicit assumption: nonhuman animals, nonhuman forests, and nonhuman streams were commodities that could be bought and sold. An animal was little more than an instrument to be used.

But Anishinaabeg peoples in the Great Lakes saw sturgeon as relatives and fellow spiritual beings, just as Native American peoples on the Great Plains saw bison as sacred relatives. Sturgeon ghosts swam in the empty streams, in the stories and songs of those people who mourned them. The Native American writer N. Scott Momaday said of the destruction of the bison in this same time period: "It would be hard to imagine anything more deeply hurtful than the loss of something ineffably sacred. . . . Why is this happening? . . . this wholesale slaughter must have been first confusing, and then—you know—a devastation. A wound in the heart that we cannot conceive of now."[52] White Horse, an Omaha man, spoke for many Indigenous peoples when he said of the wholesale slaughter of their nonhuman relatives: "Now the face of all the land is changed and sad. The living creatures are gone. I see the land desolate, and I suffer unspeakable sadness. Sometimes I wake in the night and feel as though I should suffocate from the pressure of this awful feeling of loneliness."[53]

Lake sturgeon were to the upper Great Lakes tribes what salmon were to the Pacific Northwest tribes. Both fish offered material sustenance and resilience to the human cultures who patterned their own seasonal movements after their fish relatives. But sturgeon, like salmon, were far more than material resources: they offered cultural and spiritual sustenance as well. Sturgeon tied together the watersheds of the north, and their collapse indicates some deep flaws in the ways Euro-Americans relate to and transform nature.

For Pacific Northwest tribes that long depended upon salmon, it is unthinkable that Euro-Americans could have destroyed so many salmon runs so quickly, in just a few human generations. Restoration efforts remain bogged down in politics and salmon runs continue to dwindle. But for the Pacific Northwest salmon tribes, a future without those salmon is no future at all. Your familial connections to place, to the future, to the ancestors, are all severed. If sturgeon were to be eradicated, Great Lakes tribes would face similar losses.

Historian Richard White wrote this about salmon peoples in the Pacific Northwest:

> For thousands of years Indian people had recognized and understood the blessing of a world in which small fish left the river, harvested the greater solar energy available in the ocean, and returned as very big fish. These fish always returned at the same time to the same place, and . . . . Indians knew these places and had developed techniques that allowed them to expend less energy on capturing fish than the whites who followed. . . . "A greater One than we had done all the planting, and

fisherman had only to reap," said one Indian fisherman. It was as if seed wheat left home in April to return as a field of grain in September. It was as if deer came walking through town every November.

And in the face of such regularity and bounty, the Americans began breeding the fish in factories and setting out to sea to catch them. Each step of the process that led to this result was logical. It was only the result that was mad. Like many kinds of madness, this one looked quite sane from the inside. One thing followed quite understandably from another until both a kind of environmental insanity and a bitter social conflict were achieved.[54]

Salmon, like sturgeon, persisted across enormous natural variability: volcanoes, earthquakes, ice ages, continents plowing back and forth into each other. Salmon returned, in the stories of Pacific Northwest tribes, "as long as native fishermen felt reverence for the Salmon People; and they gradually stopped returning after science declared the earth a vast storehouse stocked with uniform and interchangeable salmon-parts."[55]

Similarly, for Indigenous peoples in the upper Great Lakes, lake sturgeon aren't just commodities—fungible economic units. Sturgeon live in relationship, with each other and with people. They are (as David James Duncan writes about salmon) "transrational beings whose living bodies bring far-reaching blessings to a watershed. . . . Their existence puts us in touch with ultimate questions, their annihilation with ultimate consequences." Sturgeon, like salmon, are a "spiritual gift, so their vanishing is spiritual loss." In 1999, the Northwest bishops of the Catholic Church joined with the tribes in formally declaring the Columbia and Snake Rivers a "sacred commons" created by God to be shared and lovingly cared for by all. The bishops condemned current "arbitrary policies and practices based primarily on the greed and politics of power" and called for holistic solutions. The bishops and the tribes agreed: "it is not possible for individuals or governments to comprehend, analyze, or defend a living holiness from a purely quantitative point of view."[56]

Migration is core to what the bishops called the "living holiness" of salmon, and the same is true for sturgeon. Sturgeon, like salmon, were so abundant that their migrations brought the nutrients of open waters high into headwater streams, enabling entire cultures and ecosystems to blossom off their bodies. Sturgeon deaths were offerings for enriching the larger watershed (although this is even more true for Pacific salmon, who die after spawning).

Migration can bring abundance, but it also can bring unpredictability, because sturgeon, like caribou, don't always return to exactly the same place at exactly the same time. Part of the "living holiness" of migratory fish is their wild unpredictability, their escape from purely human control. Sturgeon, like all migratory creatures, aren't evenly distributed in space or time. Their abundances are threaded in rivers that run across the watershed, with levels that can fluctuate, depending on the depth of the snowpack, the quantity of rain and flood, the signals that draw sturgeon home. Their abundance is threaded in time as well as space, because you never know precisely when those signals will occur on any particular river in any particular year.

Because migrating individuals aren't entirely predictable, Indigenous peoples have learned to pay close attention. There are endless stories of northern peoples watching closely for signs that the caribou, sturgeon, or seabird migrations would arrive, saving the people once more from famine. Such stories helped to pass this intimate knowledge of the natural world from one generation to another. When sturgeon, like salmon, fade from our rivers, we have to be careful that they don't also fade from our memories and stories, because then we will no longer recognize the value of the efforts that we need to make for them. Restoration takes a lot of heart, and you sustain that heart with stories, not just with money.

Indigenous Restoration

Restoration for sturgeon changed when tribes in the Great Lakes region won federal recognition of their reserved treaty rights to co-manage fisheries on ceded territories. In the nineteenth century, the U.S. government made three major treaties with the Anishinaabeg in the Lake Superior basin, establishing reservations under tribal control. The Anishinaabeg were careful to retain the right to hunt, fish, and gather on ceded territories, which also meant the right to participate in management of natural resources.[57]

State governments refused to recognize these ceded territory rights until court cases in the late twentieth century forced them to do so. For decades, Wisconsin had arrested Anishinaabeg who fished and hunted on ceded territories without state licenses, which the tribes insisted were unnecessary for them. In 1974, two members of the Lac Courte Oreilles band were arrested for spearing fish on ceded territories. The band sued the state. In 1983, a federal court upheld off-reservation treaty rights in a landmark judgment called the Voigt Decision. After the Voigt Decision, state agencies were initially reluctant to give up their authority.[58] Eventually, however, most agencies became willing to partner with the tribes.

Once treaty rights to manage fisheries on ceded territories had been recognized, the tribes could lead restoration of sturgeon, bringing back their ancient clan partners to the waters of the north. The sturgeon restoration programs led by the Little River Band of Ottawa Indians in the Big Manistee River of Michigan offer an excellent example. As Holtgren and colleagues explain, "By the early 2000s, only 40 to 50 fish a year returned to spawn in the Big Manistee River. Many historic nmé rivers lost their populations completely." Hatchery programs had brought a few nmé back, but they "came back to the river, not as a healthy component of either the river or tribal culture, but weakened, embattled, and imperiled. Today, nmé are at less than 1 percent of their historic numbers and are listed as either threatened or endangered by 19 of the 20 states within their original range. In many ways, especially to settler Americans, nmé became a forgotten fish."[59]

Some Euro-American biologists and policymakers were impatient with efforts to restore individual runs of lake sturgeon to specific reaches of river and stream (just as they're impatient with efforts to save individual populations of woodland caribou). The Endangered Species Act focuses on populations, not individuals. Some historic runs contain only a few individuals, barely hanging onto survival. So why bother with them? Yes, save the species, as long as the law requires that. But for individual runs, wouldn't triage make more sense, just as in a coronavirus ward, doctors and nurses reserve respirators for those who are most likely to survive?

From Indigenous perspectives, even small communities of sturgeon are worth restoring. What matters is not just sustaining individual runs of migratory fish in the watershed. What matters equally as much is sustaining cultural memories and practices, because without those, restoration will become impossible.

The Little River Band insisted on leading an unusual restoration effort shaped by an Indigenous conception of stewardship. They "focused not just on habitat or reproduction, the typical currency of fisheries biologists, but also on revising the band's 'cultural and ecological connection to their nonhuman kin.'" Initially, they met with broad resistance. Some white folks were furious, feeling that treaty rights "were unfair to settler Americans." The Michigan Department of Natural Resources resisted as well, arguing that the tribe lacked legal authority to engage in fisheries restoration. Other whites thought sturgeon were too far gone, figuring "it was a waste of time and money to restore a fish that might take 100 years to recover fully."[60]

But the tribe persisted, for they were "convinced that by reclaiming the nmé's rightful place within the watershed, balance would be restored"—not just for the fish, not just for their own members, but for "the river's other nonhuman kin." A framework of kinship, of "restoring relatives," prevailed in the stewardship program. The tribe launched the 7 Generation Program in 2001; a year later, their biologists found evidence that lake sturgeon were naturally reproducing in the Big Manistee River. This sign that sturgeon were not just ghosts, but were persisting in spite of the myriad threats to them, helped persuade skeptics that the project was worthwhile.[61]

In 2004 the tribe "established the first portable streamside rearing facility where young nmé collected from the river in spring were nurtured and protected until the fall, when they were released back into the same river. This was the first time the technique had ever been used for lake sturgeon. Its success was especially important because this technique kept the fish 'home' during rearing and because maintaining the genetic makeup of nmé populations in different rivers is crucial for the Anishinaabeg, whose traditional beliefs include specific relationships between particular families and their local fish populations. This idea also supports the principles of conservation biology, which seeks to maintain the unique genetic attributes of each river's nmé population."[62] "Since 2004, another five streamside facilities based on the same design have been set up within the Lake Michigan Basin. The tribe's Nmé Stewardship Plan is now a guiding document for nmé restoration in the Great Lakes, and it has changed how the region's fisheries are managed today. Many agencies, including the U.S. Fish and Wildlife Service and the states of Michigan and Wisconsin, now collaborate with the tribe's sturgeon program."[63]

In addition to habitat restoration, the tribe also includes active interventions to help protect young fry from threats. For example, to control invasive sea lamprey in the Lake Superior basin, state and tribal agencies must use a powerful chemical called lampricide. While harmless to adult native fish, the chemical can harm young sturgeon. Therefore, tribal biologists collaborate with staff from state agencies, netting young sturgeon in the river and transferring them to one of the streamside rearing facilities, where they can swim in local stream water free of lampricide for up to ten days. They are then released back into the main river, better able to survive.

Fisheries biologist Marty Holtgren explains: "Once they get to a certain size, the survival rate is high, maybe 50 percent." The program's success is visible not only in sturgeon numbers, but also "in the changed relationship between the settler Americans in the Manistee area and the Anishinaabeg. Fishing guides, who previously were suspicious of the restoration initiatives, have begun voluntarily helping locate sturgeon. Outfitters educate their clients about the importance of nmé and tribal stewardship. Settler Americans are now happy to see the tribal biologists on the river. . . . The watershed community now views nmé as a species whose presence makes the Manistee area special. The fish has been able to both heal old wounds and create new, sustainable, relationships among people, even in a watershed where these relationships have been strained by settler colonialism. Now, every September hundreds of people—Native as well as settler—gather by the banks of the Big Manistee River to release young nmé into the waters with much fanfare, ceremony, and feasting. Each person cups a young sturgeon in his or her hand and gently guides it back to the river [see plate 8]. A lasting connection is built between fish and human. Most non-Natives present at the ceremony are probably not ready to adopt the Anishinaabeg way of thinking about the world and our place in it. But in that moment of release they embrace a sense of themselves linked to a watershed shared with their nonhuman kin."[64]

In Minnesota, near the Red Lake refuge where caribou restoration had failed, tribal communities now collaborate with federal and state biologists on sturgeon restoration programs. Hatchery programs are still a critical part of their restoration efforts, but the White Earth tribe also focuses on restoring watershed connections. Program managers write that "Lake sturgeon like to spawn on rocky substrate which is often found in the rivers. With all of the dams we have dotted across the landscape in the Midwest, lake sturgeon are up against fragmented habitat—they are essentially separated from their lake habitat and their river spawning habitat."[65]

The tribe and the Minnesota Department of Natural Resources have collaborated on more than thirty fish-passage projects. At certain historic dams on the reservation, they alter "the dam face by putting rocks and boulders immediately downstream of the dam, in turn creating a gently sloping rock-riffle run, allowing fish to pass through. Fish passage projects like this aim to restore the connectivity of the river, allowing lake sturgeon to access the habitat they need to spawn, and ultimately reproduce on their own." The long-term goal is "to facilitate the development of self-sustaining populations of lake sturgeon, and provide the tools necessary to restore connectivity to their habitat and boost their populations where they once thrived."[66]

Pollution control is another focus of tribal restoration programs. The Fond du Lac Reservation is in the St. Louis River watershed, near the western end of Lake Superior. Iron mining, paper pollution, oil development, and the U.S. Steel Corporation concentrated toxics in this watershed and, together with commercial fishing, the combined stressors brought about the demise of lake sturgeon here by about 1900. "It wasn't until 1979, after the Clean Water Act, that the Western Lake Superior Sanitary District began treating domestic and industrial wastewater before it entered the river. By the early 1980s, water quality had improved enough to attempt restoring lake sturgeon." This isn't a fast process, but hatchery sturgeon in the St. Louis watershed have begun to spawn (see plate 9). New surveillance technologies have assisted this work. Acoustic transmitters implanted in individual fish can track their movements, "sending out signals that are picked up by receivers in the river and in the estuary near Duluth." Tracking allows biologists to figure out how individual fish are using different areas within the estuary, so they can focus on the most important sites when it comes time to clean up industrial wastes such as PCBs.[67]

PLATE 1 One of the last of the woodland caribou in the southern
Selkirk Mountains of Idaho, photo taken on October 22, 2007.
Steve Forrest, USFWS Pacific Region

PLATE 2 Woodland caribou on Michipicoten Island in Lake Superior.
Christian Schroeder

PLATE 3 Woodland caribou inhabit the boreal forest, which is at increasing risk with climate change. *Christian Schroeder*

PLATE 4 The Sámi have herded reindeer in the boreal north for centuries, but climate change, mining, and industrial development threaten their ability to continue herding. *Mats Andersson / Creative Commons*

PLATE 5 Lake sturgeon can live for over a century and reach enormous sizes. *Todd Stailey / Tennessee Aquarium*

PLATE 6 Lake sturgeon have a tube-like mouth that allows them to suck up food from lakebeds. *Todd Marsee / Michigan Sea Grant*

PLATE 7 Derek Harper, *Sturgeon and Shadow*, 2012. Native American tribes honor the lake sturgeon as a sacred relative, calling the fish *nmé* or grandfather.

PLATE 8 Young-of-year lake sturgeon captured in a bottom trawl in the lower St. Clair River as part of a monitoring and restoration program, September 2011. *USFWS*

PLATE 9 This lake sturgeon was discovered during fish sampling on the Bad River in northern Wisconsin. *Sharon Rayford, USFWS*

PLATE 10 Common loons have come to represent the soul of lake country for many North Americans. *Richard Simonsen*

PLATE 11 Sunset over Seney National Wildlife Refuge.
R. L. Drieslein, USFWS, Division of Public Affairs

PLATE 12 Loons eat fish, which makes them vulnerable to mercury and other toxics that accumulate in their prey. *Bert de Tilly / Creative Commons*

PLATE 13 Loon chicks often ride on the adult's back, which protects them against chilling and exposure. Adults poisoned by mercury offer less parental care to the chick, reducing survival. *USFS, Northern Region, Missoula, MT*

PLATE 14 In the 1950s, the state of Michigan hoped to draw visitors from around the region to vacation in the lake country. *Michigan Tourist Council*

PLATE 15 Common loons, like other migratory birds, are now part of our climate fingerprint on the world, as they track north, trying to find cooler, cleaner places to continue their ancient migrations. *Alan Schmierer / Creative Commons*

Menominee Restoration

While so far this chapter has focused on Anishinaabeg peoples, other Great Lakes tribes such as the Menominee have equally close relations with sturgeon. Like the Anishinaabeg, the Sturgeon Clan plays a critical role in Menominee spiritual and community life. "The creation of all the clans began when the Good Spirit who had created the earth granted a bear the power to change forms. The bear became a human. As Bear was lonely, he invited other animals to join him, one of which was a sturgeon. They all became people and founded the clans of the Menominee tribe."[68] Lake sturgeon were important in the material lives of the Menominee as well as in their spiritual lives. For many generations, tribal members ate fresh fish, smoked fish for storage and, like the Anishinaabeg, harvested isinglass from the swim bladders to use as glue and to sell to traders.[69]

To honor the importance of sturgeon in Menominee culture, the tribe has for centuries held a festival during the spring spawning runs. The festival recognizes "the importance of the sturgeon in founding their clans. . . . One of the important aspects of the Sturgeon Festival is the Fish Dance, which is done to ensure a good harvest and a good spawning. The fish are considered to be the spiritual protectors of the wild rice; a food which is also important to the tribe."[70]

In the nineteenth century, the federal government ordered the Menominee to select a small portion of their ancestral territory for a reservation, seizing the rest for white settlement. The Menominee selected a site on the Wolf River, partly because of the annual spring run of sturgeon "below Keshena Falls. The annual run of sturgeon provided much needed sustenance each spring to the tribe after the long Wisconsin winters depleted food stores by late April when the sturgeon would return to the Falls."[71]

The sturgeon that spawn in the Wolf River spend much of their lives in massive, shallow Lake Winnebago. A series of dams built in the late 1800s blocked sturgeon migration into the upper Wolf River and the Menominee Reservation. State, public, and tribal efforts several decades ago restored a small portion of the migration up to Lake Winnebago, and watching those spawning sturgeon had become a very popular activity in the small towns—so popular that poachers targeted the fish at night. The tribe worked with local fishing clubs to send out guards at night to welcome the fish home and protect them from poachers.

Sturgeon for Tomorrow, a predominantly Euro-American group of recreational fishing enthusiasts, had also worked hard to restore spawning habitat and find funding for hatchery projects. But the old dams still prevented the lake sturgeon from migrating upstream to Keshena Falls, a sacred site within the Menominee Reservation itself.

Although many whites had forgotten the historic abundance of sturgeon that once migrated up to the falls, Menominee elders had not. And the fish probably had not either, because, according to Ron Bruchs and Ryan Koenigs of the Wisconsin Department of Natural Resources, "very likely there are still some sturgeon in the Winnebago population who were hatched at Keshena Falls prior to the construction of the dams at Shawano in the late 1800s."[72] These fish live a very long time, and the Menominee believe that the sturgeons' bodies bear the memories of these distant migrations.

In 2011, the Menominee tribe and the Wisconsin Department of Natural Resources agreed to try a risky effort, physically transferring one hundred adult sturgeon each year. They would catch them, remove them from the river, then truck them around the dam, to see if they might continue their migration upstream onto the reservation. Sturgeon for Tomorrow and other Anglo recreational fishing clubs tried to block the project, fearing this might imperil too many adult sturgeon. Moreover, the white folks thought it was biologically pointless. Why did sturgeon need to swim all the way home to the reservation, when they had figured out other places to spawn?

Sturgeon for Tomorrow sued the tribe and lost in court, allowing the project to move forward. The tribe and state decided to implant each individual fish with a sonic transmitter, allowing them to track each fish's movements. This let them determine "if fish stayed in the river to spawn at Keshena Falls, took up residence above the dams, or moved back downstream to where they were originally captured. . . . As the spring spawning migration grew near following the transfer of the first 100 fish, Walter Cox, Director of the Menominee Conservation Department, and Craig Corn, Menominee Tribal Chairman, worked with DNR Law Enforcement Supervisor Carl Mesman and his staff to set up a Sturgeon Guard program at Keshena Falls to watch for fish if and when they showed up to spawn. . . . Menominee sturgeon guards worked around the clock, and tribal elders silently watched day in and day out for the return of the fish. Ron Bruch and his crew joined Don Reiter and Richard Anamita, the Menominee Tribal fisheries biologists, each day searching, waiting and hoping."[73]

Much to the delight of tribal members and fisheries staff, the individual transmitters showed that translocated adults were not only surviving—they were also spawning at the falls. The Lake Winnebago–Wolf River–Fox River system is now home to the largest remaining population of lake sturgeon in the world, revealing how important these restoration efforts have been to the future of the species. This wasn't just a scientific breakthrough; it was also a cultural breakthrough. When the sturgeon returned to spawn below the sacred Keshena Falls, after a century of absence, "All those seeing this grand spectacle that had not been seen for over 100 years were moved and deeply felt the sense of historical and cultural significance of this event."[74]

Part of what makes these creative restoration projects possible is new tracking technology. In the latter half of the twentieth century, with the development of military technologies such as radio-telemetry, GPS-enabled data-loggers, and remote sensing, the ability to track and visualize population movements was transformed. Scientific understandings of migration had initially relied on technologies that focused on population-level movements. The new technologies allowed individual animals to be tracked, making the formerly invisible world newly visible. If proper respect is paid to the individual fish, these new tools fit well with the practices of Indigenous peoples in the north, who for centuries had developed ways of understanding migration based on close observations of individuals. As anthropologist Damian Castro reports, Indigenous peoples recognize in individual elder animals "a sentient, social being whose level of autonomy is comparable to the humans."[75] In caribou, for example, Indigenous people recognized that certain individuals lead group migrations, so you need to protect those specific animals so the whole group can find its way home.[76] In whooping cranes, a group's most experienced member shapes the "navigational performance of a group." When the elder cranes die, restorationists must "employ ultra-light aircraft to teach migration routes to inexperienced migrants, in the hope that they will in time become tutors themselves."[77]

Knowing Fish

The Indigenous restoration projects described here require a much more involved connection with individual sturgeon than some wildlife biologists may be comfortable with. What's wild about a sturgeon you rear and place above a dam? But these active restoration projects may be what we increasingly need to do in the Anthropocene: come up with new ways of thinking about wildness that acknowledge continuing entanglements between human and nonhuman. Perhaps it's these renewed connections that are now necessary to sustain migratory species in the north.

How do we make those entangled connections with sturgeon who, like all fish, are profoundly other to humans? They swim through the world, sensing and acting and living and dying, in ways far different

STURGEON CAUGHT AT BIG FALLS MINN
JUNE 1 '08. WEIGHT 112½ lbs.

FIGURE 4.3 Two girls with a lake sturgeon caught in Koochiching County, Minnesota, near the Big Bog, 1908. *Minnesota Historical Society*

than we do. They breathe in a watery medium that would quickly choke us. They see, not with the same eyes we see, but with electroreceptors on their heads that respond to electric signals, and they use these to feed, mate, migrate, and hang out with their pals. They communicate, but not in a language we understand.

One way many cultures have engaged with individual nonhuman animals is through anthropomorphism, projecting human attributes such as love and generosity onto other animals. Scientists warn against this, telling us that it only blinds us to the complexities of other animals. By letting us rest in our assumption that other animals are just imperfect imitations of ourselves, it can blind us to the complexities of their own life histories. On the other hand, to refuse anthropomorphism can be equally treacherous—if we follow Descartes in assuming other animals are merely machines, without culture, without faith, without feelings.

How do we navigate this dilemma, avoiding the risks of anthropomorphism while rejecting reductive, mechanistic thinking? Western biologists try to observe the nonhuman world as closely as they can, then move from these empirical observations into tests of theory, modelling population costs and benefits over evolutionary time. Indigenous communities also start with close observation of the nonhuman world, trying to be as accurate as possible about the specific patterns and processes they observe. But then, rather than assume that only populations matter in evolutionary history, they also pay attention to individuals. In Indigenous perspectives, sturgeon—like humans—have culture, religion, complexity, and meaning in their individual lives, just as humans have.[78] Stories and rituals teach children to pay attention, to see the world outside of human boundaries, to grasp what the Fond du Lac call *Giga-waabamin*: a promise, in essence, that "I will see you."[79]

This is what Indigenous scientists now call "two-eyed seeing." Two-eyed seeing integrates Indigenous traditional knowledge with western scientific research. With two-eyed seeing, restoration becomes more than simply removing invasive species and restoring natives. Rather, restoration becomes "cultural renewal, a journey from the mind to the heart."[80] Such two-eyed seeing may be essential for a sustainable future in our increasingly precarious but precious world.

FIVE

The Gift of the Loon

Most summer mornings, I drink my morning coffee while perched atop the sandstone cliffs overlooking Lake Superior. If I'm lucky, I'll hear the tremolo of a common loon (*Gavia immer*). Lake Superior is too rough for loons to breed except in a few sheltered spots, but young males often gather on the lake and practice their calls. It turns out loons have a lot to learn—their calls are not purely instinctive, but rather are cultural practices learned across generations. Young loons starting out sound like throttled wolves, shrieking and howling with enthusiasm but little skill.

One morning, twenty-eight loons swam into view through the fog, yipping like puppies—a gift I'm unlikely to forget. Every time I hear a loon, no matter how much practice they still need to perfect their calls, I am filled with hope, for their calls suggest something wild, outside the bounds of human constraint.[1]

I'm not the only cottager who feels this way, of course. Common loons have come to represent the soul of northern lake country (see plate 10). As journalist Sharon Guynup writes for *National Geographic*, "For many North Americans, loons are a much-beloved bird, symbolizing the solitude of the deep-woods wilderness with their distinctive, haunting wail that echoes over the northern lakes where they breed in summertime."[2] These are powerful beliefs, part of the cultural identity of many Americans who vacation in the north woods.

Cabin culture in the north woods is replete with loon images. You can stop at any number of gift shops on your way north from Chicago and snag a loon sign. How about a loon pepper grinder? Or a loon comforter for your log bed? Mercer, Wisconsin—Gateway to the North! (and former Ku Klux Klan conclave, where a chapter was active as recently as 2013)—has a giant loon statue welcoming visitors.[3] If so inclined, you can buy yourself a loon toilet seat.

Canadians share this love for loons (and loon kitsch). In Canada, even the one-dollar coin is festooned with a loon. Yet the cultural meanings of loons have been remarkably shape-changing. These cultural identities have a complex history, and that history influences the ways we manage the loon's future.

Like the woodland caribou and lake sturgeon, the common loon is a migratory species that breeds in the boreal north. According to the National Park Service, they have been "on earth for about 70 million years, making them one of the most ancient bird families. Mammoths, mastodons, and saber toothed tigers heard the loon's voice ringing out" over watersheds that have undergone tremendous transformation, with the advance and retreat of glaciers, the warming and cooling of climates, the rising and leveling of mountain ranges.[4] Yet after persisting through these changes, their future is now threatened by humans. Loons have been the focus of decades of restoration efforts—yet they remain at risk in northern landscapes.

One of five loon species across the global north, the common loon is the most abundant in North America, numbering about 258,000 breeding pairs, with about 94 percent of those in Canada.[5] These numbers seem high, suggesting that loons aren't yet on the verge of extinction. But the combination of toxics and climate change mean that they could easily be extirpated in the United States and southern Canada. As the Minnesota *Star Tribune* warns its readers: "Minnesota could lose its beloved state bird in coming decades if humans don't stall climate change and prevent the common loon from shifting north." The article goes on to tell us that loons "define Minnesota as much as lakes, snow and hot dish"—but they are "among 55 species likely to disappear from the state for the summer by 2080" if climate change continues at its current rate.[6]

While only 182 bird species have gone completely extinct—that we know of—a recent report by the National Audubon Society found that nearly two-thirds of North American species (389 out of 604) are vulnerable to extinction if the climate warms by 3 degrees Celsius. At 2 degrees of warming, over half could crash.[7] Billions of birds are already missing—as a study from Cornell University warned us in *Science*.[8] Waterbirds, shorebirds, grassland birds, and long-distance migrants are all declining. Why? If we love birds as much as the $5 billion spent on bird food in 2018 suggests, why can't we stem their decline?[9]

This chapter explores tensions among ideas of wild nature, the history of loon restoration, toxics, energy extraction, and avian extinction.[10] Euro-American beliefs about an association between loons and remote wilderness have influenced the ways Euro-Americans have treated loons. Initially, that association led to a campaign to exterminate loons. Now it leads to the opposite: efforts to protect loon nests and restore habitat. Yet an association between loons and wilderness obscures the ways that urban decisions about toxic substances, fossil fuel combustion, and extractive industries affect loon populations breeding thousands of miles away.

Cultural Beliefs

Across the north, loons are extraordinarily important in stories, religion, and teaching. Indigenous peoples "know the loon intimately, not

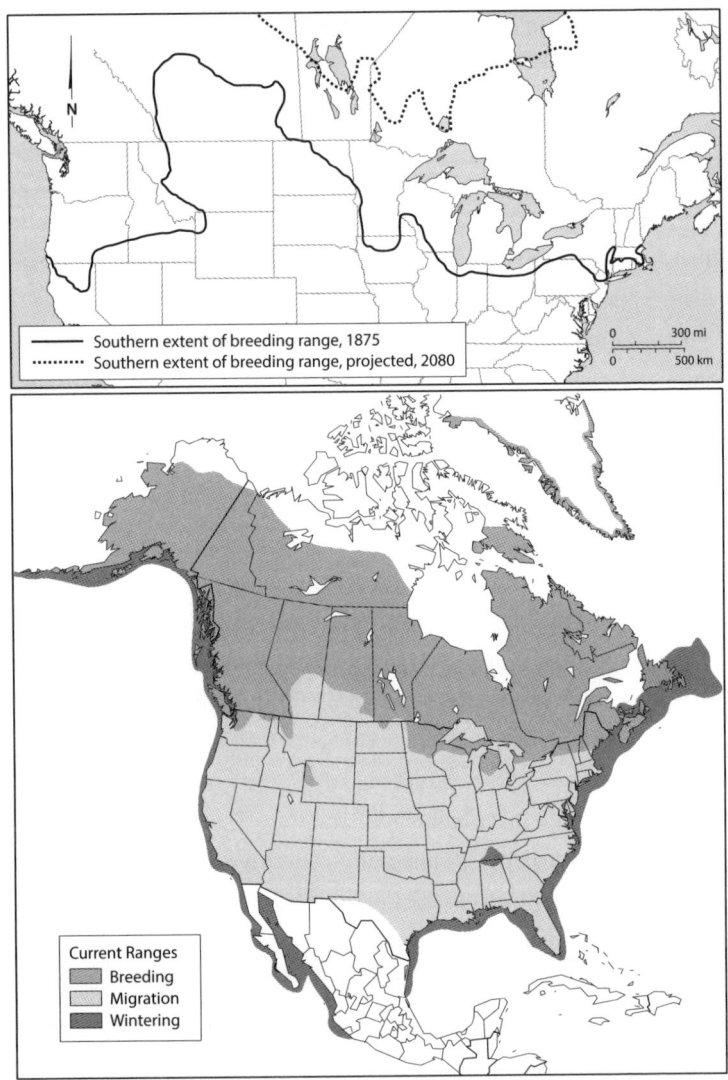

MAP 5.1 The breeding range of common loons has decreased across North America and, with climate change, is expected to retreat even further by 2080. *Bill Nelson*

only as a physical presence but also as a powerful spiritual being."[11] For many northern Indigenous peoples, loons can communicate across the boundaries between human and nonhuman animals, with links to what anthropologist Richard Nelson calls "distant time" when human and nonhuman shape-shifted at will.[12]

In one story, shared in different forms across Inuit cultures in the north, a cruel mother steals the sight of her young son because she's jealous of his hunting prowess. A loon hears him crying by the lakeshore, and she swims up and asks him what has made him so sad. The boy tells her, and she tells him to climb onto her back, hold on tight, and keep his eyes shut until she tells him otherwise. The boy must trust the loon even though he can't trust his own mother.

He obeys. She dives very deep three times, and when she surfaces the final time, she tells him to open his eyes. He can see! She has lent the boy her powerful sight, so the boy says to her, "I must give you something in return. What would you like?"

"Only to know the lake will always be stocked with fish," the loon says. "Only to know you will guard all the gifts of this world."

"I will look after you, always," the boy promises. "I shall treasure this world and all I see."[13]

As this story suggests, loons were associated with vision and perception across many northern cultures, possibly because of their bright red eyes during the breeding season. Loons assisted shamans across northern Asia and North America. Anthropologist Emily Auger writes that "the ability to dive vast depths under water, with keen vision in both air and water . . . made the loon an ideal guide for the shaman on his spirit journeys."[14] For both the Inuit and the Koyukon, a Dene people living along the Yukon River in Alaska, the loon was "associated with superhuman and ceremonial greatness." Like the Inuit, the Koyukon tell of how a loon restored the sight of a blind person. Nelson writes that Distant Time stories tell of how "the man who became dodzina, the common loon, used his medicine to restore another man's sight."[15] For many Koyukon, the trailing stars of the Milky Way represent the flight of the loon.[16]

Loons are one of the key shape-shifting creatures for the Copper Inuit, whose "representation derives from earlier times, when men and animals could take each other's form at will by pushing a muzzle or beak cap up or down." Masks made from the yellow-billed loons confer this power, if the beak cap is pushed down by a shaman during a ceremony.[17]

Northern peoples believe loons to have special connections with the human voice as well as vision. For the Koyukon, Distant Time stories tell of how the Koyukon people's songs "originated partly from the loon's crying and partly from the human imagination."[18] Knud Rasmussen, the Greenlandic Inuit-Danish anthropologist, noted that, for the Inuit, the loon was "the bird of song,", and so a parent might place the bill of a loon in a young child's mouth to lend the gift of song.[19] "The Caribou Inuit told Knud Rasmussen . . . a story about how a loon, upon hearing a human weeping in grief, tried to imitate the sound and, since then, loons have cried like humans"—suggesting deep connections across the two species.[20] The Cree of northern Canada believed the loon's call was the cry of slain warriors calling back to the land of the living.

Some northern cultures hunted the common loon, luring the flocks by imitating their calls and shooting the birds from behind a blind or from a boat. Many Indigenous peoples, however, avoided eating loons. Because some tribes perceived that loons can shape-shift with people, eating them was taboo, verging on cannibalism. The peoples of the northwest coast of North America, particularly the Eyak of southeast Alaska, refused to hunt loons because their stories described a boy turning into a loon. Instead, they observed loons closely, believing that when loons became restless, storms were coming. The Koyukon feared that by eating loons, they might take on the loon's clumsiness on land, becoming "awkward and slow."[21]

In Anishinaabeg belief systems, the loon was not just another bird, but rather "the first act of creation, when the very voice of the Creator echoed across the void and became embodied in a gray and black shadow, the spirit of the loon."[22] Like the Inuit, the Anishinaabeg also have a loon story; similarly, it's about the loon's vision and protectiveness. The Loon came upon Nanabush in the process of tricking a group of ducks into competing for an invitation to a dance party, asking the ducks to dance past him, eyes closed. In truth Nanabush planned to kill the ducks for his dinner, but Loon, observing, called out to warn the ducks. In retaliation for ruining his plans, Nanabush gave Loon a sharp kick, knocking his legs so far back that Loon would never be able to walk on land again, but would be confined to water.[23]

As migratory people followed loons, caribou, and sturgeon north into a warming world, they left artistic records of their journeys. "The very first accounts of the chemistry between loons and people . . . can be seen in the granite outcroppings of the Canadian Shield, north of the Great Lakes," where pictographs of loons are still visible.[24]

In northern European cultures, loons had complex cultural meanings as well, but those meanings tended to be far more fearful. In one medieval Icelandic saga, Gigsli, the doomed outlaw hero, has a dream vision of his impending death. He dreams of two giant loons that fight one another, covered in blood: "I, maker of the sword's voice / heard two loon birds fighting / and I knew that soon the dew of bows would be descending."[25]

An association between loons and death, marked by what Europeans heard as "demonic laughter," developed in European cultures, leading to their persecution. Like wolves, loons symbolized a dangerous wilderness outside the bounds of civilization.[26] Europeans imagined that both loons and wolves possessed a liminal ability to cross borders between human and nonhuman, life and death, night and dusk. When Europeans invaded North America, they brought with them these beliefs. Loons were demonic, crazy, wolf-life, beyond human control—in a word, *loony*—and therefore to be eradicated.

More material fears shaped Euro-American perceptions of loons as well. Because loons also hunt fish, hungry Europeans saw them as competitors in a zero-sum game for food. Just as cormorants today are still murdered by fishermen angry about competition, so were loons until well into the twentieth century.

Henry David Thoreau built upon the association of loons with wolves and wildness. In *Walden* (1854), he described a loon's call: "His usual note was this demoniac laughter, yet somewhat like that of a water-fowl; but occasionally, when he had balked me most success-fully and come up a long way off, he uttered a long-drawn unearthly howl, probably more like that of a wolf than any bird; as when a beast puts his muzzle to the ground and deliberately howls. This was his looning,—perhaps the wildest sound that is ever heard here, making the woods ring far and wide."[27]

As Thoreau tried to move closer, the loons frustrated his efforts, and Thoreau portrayed his loon as mocking the limited abilities of humans. But unlike many other Euro-Americans, Thoreau admired the loon's "cunning," his ability to dive, and his ability to outwit hu-man hunters. He wrote: "At length, having come up fifty rods off, he uttered one of those prolonged howls, as if calling on the god of loons to aid him, and immediately there came a wind from the east and rippled the surface, and filled the whole air with misty rain, and I was impressed as if it were the prayer of the loon answered, and his god was angry with me; and so I left him disappearing far away on the tumultuous surface."[28] Thoreau's description of loons marks one of the first Euro-American texts that portrayed the loon's wildness as a mark of wild virtue. It would take over a century for most Americans to share his view.

Loon Biology

On April 14, 2020, the Seney Wildlife Refuge staff posted a joyful Facebook welcome to two of the oldest loons in the world, ABJ and Fe, when they returned from their winter homes and spring migration:

> In the coming weeks the venerable F Pool pair will vigorously defend their territory from other adult loons in search of their own breeding turf and mate. If successful in parrying these challenges, the ABJ and Fe will settle in for a record 24th consecutive season of nesting at Seney, with one or two chicks hopefully hatching in June or early July.[29]

ABJ, the male loon, had been banded on the refuge back in 1987 as a chick, making him thirty-three years old. His mate, Fe, was even older, though her exact age is unknown.

On July 1, 2020, Seney staff added a birth announcement update.

> Birth Announcement: For those of you who have been following ABJ and Fe's story I would like to announce that they hatched their first chick of the season on 6/30/2020. According to Teresa McGill, an avid Seney photographer, she witnessed the pair cooing to the chick just after it hatched. Fe, the mother, lured the chick off the nest and into the pool with a small fish.

Over 2,400 people liked this Facebook post, delighted to see the elderly pair hatch another chick. The excitement about loons at Seney Wildlife Refuge reflects the fact that much of what biologists know about loon life span, basic reproductive biology, and toxicology has come from decades of research at Seney—a place that underwent dramatic ecological collapse in the late nineteenth and early twentieth centuries, and whose recovery as an ecosystem has brought about critical recoveries for many waterbirds (see plate 11).

Like the Red Lake refuge in Minnesota discussed in earlier chapters, Seney was shaped by many generations of Anishinaabeg land

FIGURE 5.1 During the late nineteenth and early twentieth centuries, the Upper Peninsula of Michigan was heavily logged and then burned. The cutover lands were sold to hopeful white farmers, but those farms often failed. (Photo by W. J. Beal, Bureau of Forestry, 1901–1905.) *National Archives*

use. The federal 1836 Washington Treaty forced the tribes to leave the Great Manistique Swamp, where Seney Wildlife Refuge is now located, although the Anishinaabeg retained ceded territory rights to lands that were not settled by whites.[30] The treaty opened forests to exploitation by the timber industry and, starting in the 1870s, the industry targeted the region's enormous white pines. Within two decades the area "was practically stripped of its forests. Often fires were deliberately set to clear away the wreckage of past lumbering operations and to make way for new ones."[31]

After fires ripped through what remained of the Great Manistique Swamp, a land-development company promoted the false promise of peatland drainage and farming, just as had happened in the Big Bog of Minnesota. The company sold the drained peat bogs with "extravagant promises of its productivity," but hopeful farmers could never grow enough to survive. When Euro-American farmers abandoned the land, those acres reverted to the state for unpaid taxes.[32] Michigan urged the federal government to step in and fund wetland restoration, and in 1935 Seney National Wildlife Refuge was established.

Civilian Conservation Corps workers did much of the physical labor to restore 95,531 acres of wetland, bog, and forest. They moved "thousands upon thousands of yards of sand and peat to build an intricate system of dams, dikes, and ditches designed to divert and impound water."[33] At first Canada geese—now so numerous they're considered pests in many communities—were the focus of active restoration, for they had nearly gone extinct. But soon restoration efforts at Seney expanded to a much broader suite of wildlife and plant communities.

While each annual report briefly mentioned loons, it wasn't really until 1992 that loon research intensified at Seney. Concerned by declining loon populations across the state, in 1992 Michigan drafted a loon recovery plan, required when a species is listed as threatened. But the recovery plan's authors ran up against a core problem: they lacked the basic research on loons necessary for a recovery strategy. How long do loons live? When do they begin to breed? How many return to the places where they were born in order to breed? Where do they go when they're not breeding in the northern lakes? What kills loons? Biologists didn't know the answers to any of these questions.

Researchers at Seney set out to "address these deficiencies of understanding." What they found surprised them: loons live a very long time and they return to their own particular lakes for decades. They show exceptionally high return rates to the refuge—in one year, 96 percent returned. Even more surprising, demographic patterns suggested that 40 percent of refuge loons live more than three decades.[34] That's a very old bird.

Loons arrive in pairs at Seney as soon as the ice on their lakes begins to thaw, typically in April. When we visited Seney one April soon after the ice melted, we were delighted to hear the varied yodels, tremolos, and calls of the returning loons. These vocalizations, refuge signs told us, serve many functions associated with reproduction. The tremolo can signal greeting, alarm, or warning, and different loons can send "very different messages by increasing the duration and frequency of the tremolo." Males alone yodel, using the call to defend their nesting territories. Sigurd Olson wrote of the loon yodel as "the weirdest and wildest of calls . . . beautiful and thrilling . . . maniacal and blood-curdling."[35]

The loon is a creature of water—and those waters are increasingly threatened, toxic, drained, overcommitted, and under-protected. Loons return to the same mate and same lake year after year, unless reproduction fails (see plate 12). They build their nests at the water's edge, providing quick access to the fish they eat, but making nests vulnerable to changing water levels and other disturbances. Because loons are solitary nesters, each small lake is home to only one pair of loons, who typically return to that exact same spot year after year (unless disaster or divorce intervenes). On larger lakes, breeding pairs may claim a separate bay or section, keeping well apart.

Loons are famous for sticking close to their mates. Modern Americans project a great deal onto this supposedly passionate pair-bond. In this new view, loons are no longer the demonic other; they have become a model for an oft-desired, rarely achieved human monogamy. Biologists estimate the life span of the common loon at fifteen to thirty years; for most of that time, pair stays very close to each other, or so the story goes. Recent banding studies, however, have shown that loons will sometimes switch mates after a failed nesting attempt, even between nestings in the same season. They have a divorce rate of about seven percent.

Males compete with each other, sometimes quite violently, for females. Females as well as males defend the nest from intruders. If male floaters—the young birds—have not acquired a territory by six to eight years of age, they try to "seize a territory from an established owner after a violent and prolonged territorial battle. Such battles can be dangerous. In fact, about one-third of all territorial takeovers among males result in the death of the displaced male owner."[36]

Once territories are defined and defended, things settle down a bit. The pair swim and dive quite close together, seldom straying more than twenty yards from each other. Both sexes build the nests, defend the nest site against intruders, and incubate the eggs, turning the eggs occasionally during the twenty-six to thirty-one days of incubation. Once the chicks hatch, they are precocial, meaning that they have some feathers and can immediately swim. Nevertheless, parental care remains very important.

Chicks spend most of their first week or two on their parents' backs, resting, avoiding predators, and conserving heat (see plate 13). Chilling and exhaustion kills baby loons, so close parental care is essential in these initial days.

Parents feed the chicks all the food they need during their initial three weeks. Until eight weeks of age, the adults are with their young most of the time. The chicks then begin to dive for some of their own food and, by three months of age, they're able to feed themselves and even begin to fly a bit. For the parents, this process takes a great deal of time, energy—and fish. One study found that "loons feed voraciously, mostly on small fish. A loon family—mom, dad, and chick—will eat a combined 1,500 pounds of fish in a season."[37]

Loons at Seney do remarkably well at fledging chicks, compared to most other northern areas. One refuge report notes that "Seney's long-term productivity of 0.74 fledged chicks per territorial pair is significantly higher than that of other well-monitored Michigan study sites (including Isle Royale National Park and the Ottawa National Forest), and approaches that of the highest published rates for populations across North America. This success is significantly indebted to the protection that Seney loons have always received—the refuge is functionally unique in prohibiting water-borne visitor recreation, which can often inadvertently but irrevocably disturb nesting loons."[38] Seney staff not only keep boats off the lakes; they also closely regulate water levels in the nesting pools, preventing rising waters from destroying nests, or falling waters from leaving nests stranded. All this effort establishes "a stable, dependable aquatic environment for Seney loons."[39]

While Seney staff are able to manage conditions closely enough to ensure successful fledging, most loons across the north show what wildlife biologists call "low productivity." In a world of fluctuating climate and heavy spring rains, most breeding loon pairs do not manage to fledge a single chick in a season. Because loons live for so long, a survey of adult loon numbers at any given moment might look hopeful. But if adult loons are often failing to fledge chicks, their future has been hollowed out. Sixteen years after Michigan's loon recovery plan had been written, one refuge report noted that the state still lacked the "scrupulous statewide atlasing and monitoring" necessary for understanding loon demographics.[40]

For those chicks that manage to fledge, migration is their next big challenge. Loons in the Great Lakes region spend their winters along the southern Atlantic coastline or in the Gulf of Mexico, where offshore oil production is concentrated.[41] In September, loons leave Seney before the first snows descend. Many take a feeding break near the northern edge of Lake Michigan, where they fatten up a bit, then they fly over the lake to the southern end, a long and risky journey. After resting and feeding once more, they take off for the Gulf of Mexico, flying in increments of 600 to 700 miles, at 60 to 70 miles an hour—10 long hours each day, broken up by resting and feeding, trying to time their flights to catch strong tailwinds out of the north.

Migration patterns for young birds turn out to be surprisingly complex, with the young wandering in fascinating patterns across the Great Lakes and the coasts. Map 5.2 shows the exploratory flights across much of eastern North America for one single male loon, in the years before he was old enough to breed successfully. As his early death suggests, life is hard for young loons, and many don't survive long enough to reproduce.[42]

Loon Declines

Why do loon populations continue to decline? Part of the answer lies in the past. While much beloved among birders and environmentalists now, loons have a complex conservation history. Most early conservationists felt that predators were the enemies of their efforts to restore game species deemed desirable. Therefore loons, like wolves, were targeted for eradication. For example, William H. Murray (1840–1904), who helped spark the birth of the Adirondack tourism industry, wrote of loons as demonic, cunning predators, perfect targets for a hunt. His 1869 book, *Adventures in the Wilderness; or, Camp-Life in the Adirondacks* (which went through eight printings in its first year), served as a simple guide to those who hoped to find spiritual enlightenment, physical health, and a return to man's natural state. He spent an entire chapter describing his efforts to slaughter loons on their breeding lakes.[43]

"More fish, more fun!" went the saying in the north woods, land of many lakes, and fishermen killed loons that ate the fish they wanted to catch.[44] As populations of loons and other birds declined, Congress passed the Migratory Bird Treaty Act in 1918, making it illegal to kill most migratory birds, including loons.[45] That slowed some targeted hunts of loons, although poaching continued. For example, people in Carteret County, North Carolina, kept hunting the loon, eating the fishy meat and making fishing lures from leg bones. It wasn't until a raid on May 6, 1950, apprehended nearly a hundred loon hunters that the tradition began to fade, and "a growing cultural intolerance of this practice brought the loon hunting tradition to an end."[46]

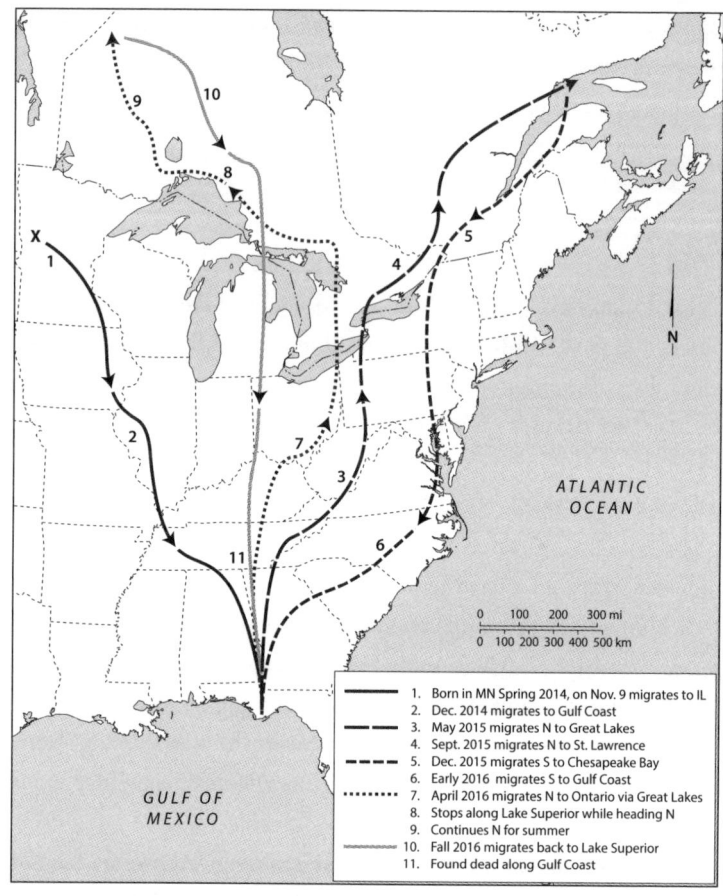

MAP 5.2 A single loon's migration path over several years. (Data from the U.S. Geological Survey's Upper Midwest Environmental Sciences Center.) *Bill Nelson*

FIGURE 5.2 Illustration from William Murray's *Adventures in the Wilderness* (1869) showing a loon hunt. *Hathitrust*

The larger threats of uncontrolled development and habitat loss weren't addressed by the Migratory Bird Treaty Act. A post–World War II economic boom, coupled with a new interest in the outdoors, led to massive lakeshore tourism and development (plate 14). Lakeshore development has had devastating consequences. As one observer noted, "Except in northern Maine and in the large federal tracts in northern Minnesota, lakes without a necklace of summer homes are hard to find these days." Because loons build nests at the water's edge, those nests are easily swamped by the wakes of powerboats and jet skis, while noise and water pollution from gasoline engines adds to the environmental stresses on adults and chicks. To help promote north woods tourism, communities attempted to tame the hordes of mosquitoes and other biting insects with pesticides. As a result, stunning quantities of arsenic, copper sulfate, and DDT were sprayed on small lakes, and those toxics entered watersheds and food chains, concentrating in fish and fish-eating birds such as loons.[47]

Even as the loon has become the symbol for the "northern experi-

FIGURE 5.3 Thomas E. Dunn shows off a loon during the John W. G. Dunn hunting and fishing expedition to Ragged and Moosehead Lakes in 1895. At that time, it was legal to hunt loons. *Collections of Maine Historical Society*

ence" in recent decades, creating that experience has threatened their lives. As one retiree observes: "People come up north from the city and hear this strange bird. This creates a lodestone. They carry the memories, often for years. Eventually they start spending more and more time up north. Pretty soon, they're living up here. Once they fall in love with loons, it's forever." For these people, "It's difficult to separate their love for loons and their love for the place loons happen to be . . . loons and loon country being one, symbol and substance, player and stage—a captivating creature in a magical land."[48] There's a familiar irony here: the love for the northwoods experience and the associated increase in development helped lead to a crash in loon populations—but also a new perception of loons, a determination to save them.

Forty years ago, local conservation organizations stepped up to protect loon nests from boaters. In Wisconsin, the Sigurd Olson Environmental Institute at Northland College started LoonWatch, a grassroots effort to "protect common loons and their aquatic habitats through education, monitoring, and research." Numerous such local efforts to protect nesting loons from powerboat disturbances and predation have helped stabilize the population in many northern lakes. Currently, only 4 percent of common loon populations are at risk from direct boating disturbance.[49] But declines continue.

Toxics, Energy, and Climate Change

Several years ago, in a Canada-wide survey, researchers realized that loon recovery had stalled. Local efforts are no longer enough to stem their decline, now that climate change, energy production, and the spread of toxic contaminants have combined their effects in the Anthropocene.[50]

Because loons are long-distance migrants, their wintering habitat and migration stopover sites are vulnerable to contamination. For example, the Deepwater Horizon oil spill released 4.9 million barrels of crude oil (about 205 million gallons) into the Gulf of Mexico. The largest risks to wildlife in the area of the spill are "petroleum toxicity, oxygen depletion, and the presence of oil dispersants. Loons eat almost exclusively fish, which will hide under oil slicks as they would under floating vegetation or sea foam. The birds become covered in oil when they surface and then ingest the oil when preening or while eating contaminated fish. Internal exposure to oil can lead to ulcers, pneumonia, liver damage, or other life-threatening conditions."[51]

Even ordinary oil production, without the catastrophe of a spill, is harmful to the environment, releasing millions of gallons of oil into marine ecosystems and taking a continued toll on water birds. This becomes the cost of doing business, so commonplace that hardly anyone bothers to count it, mark it. Another example: in 1993 the book *Polluting for Pleasure* detailed oil leaks from recreational boating amounting to the equivalent of forty Exxon Valdez spills a year.[52] By 2003, government regulations had decreased that figure to fifteen Exxon Valdez spills per year—still an astonishing amount.

Like humans, loons are vulnerable to neurological damage from lead. They die in astonishing numbers of lead poisoning—from lead pellets used by hunters and lead sinkers and jigs used by anglers. One study found that, between 1989 and 2012, nearly half of unnatural loon deaths in New Hampshire were attributed to lead ingestion. The frustrating truth is that this is needless death. Good alternatives to lead in fishing and hunting gear have long existed. The Obama administration, after years of dedicated efforts, finally passed a rule that would have outlawed the use of lead ammunition and fishing gear on federal wildlife refuges—which, after all, are supposed to be safe havens for wildlife. One of the early actions of the Trump administration was to reverse this decision.

Mercury has even more pervasive effects. Several decades ago, cottagers and scientists alike started to notice declining populations of the common loon across the upper Great Lakes. Seney became an important site for extensive research into "loon mercury exposure via fish consumption," which required understanding "temporal changes to the toxic metal's bioavailability within specific waterbodies."[53] Similar research took place on breeding lakes across Canada, and scientists realized that many loons carried surprisingly high levels of mercury in their blood and feathers. In particular, loons were being exposed to large quantities of methylmercury, the form of mercury most toxic to living things. "The mercury levels in common loons . . . are probably some of the highest levels in living animals anywhere in the world," said Mark Pokras, a veterinarian who runs the wildlife clinic at Tufts University Veterinary School in North Grafton, Massachusetts.[54] This discovery raised an alarm: pollution concentrated in northern lakes posed a significant threat to fish-eating animals—and to people.

But where is that mercury coming from? Why is it so high in places that seem most remote from industrial production? Most, if not all, mercury entering the north's surface waters comes not from local sources but rather from distant places, carried to northern lakes in the atmosphere. Large coal-burning power plants and municipal waste incinerators in the Midwest and central Canada spew mercury-laden effluent into the air. Raindrops form around these particles and wind currents carry them hundreds of miles eastward. Mercury, like many toxic substances, accumulates in living creatures. This means that, as mercury moves up the food chain, it tends to become more concentrated. Loons have no metabolic means for eliminating mercury, so its levels build up in their bodies; adult loons then pass mercury on to their eggs.

Mercury doesn't directly kill adult loons; instead it has more subtle effects, affecting reproductive success. Chicks with higher mercury levels seek energy-saving rides on the backs of adults less often, have compromised immune systems, and are less able to avoid predators. Common loon reproductive success drops to 50 percent on lakes where mercury is high.[55] Another study "found that loons with elevated levels of mercury produced one-third to one-half as many chicks as healthy loons. Mercury also leaves adult loons lethargic, which renders both adults and their chicks more vulnerable to predators."[56] As one journalist writes, mercury poisoning is an "insidious form of pollution, since it lasts for generations. . . . [E]ven chicks that make it out of the more exposed eggs alive are much less likely to survive, as they bear their parents' curse in their blood."[57]

A three-decade survey found lakes in eastern Canada are more acidic, have higher methylmercury levels, and lower reproductive success than lakes in western Canada, which are further from industrial coal contamination. But even western lakes in Canada faced challenges. Loon populations in the west are declining more steeply over time, perhaps because western lakes are warming more quickly, which increases the methylation of mercury.[58]

Under the Obama administration, efforts to reduce mercury emissions had moved forward. In April 2014, the U.S. Supreme Court upheld the Cross-State Air Pollution Rule, which would require "reductions in acid rain causing pollutants from the Midwest's power plants." The EPA planned on "smokestack emissions cuts that appear to be deep enough to halt chronic acid rain damage in the Adirondacks."[59] The Trump administration soon reversed these standards. In June 2018, the EPA under Trump "said that it doesn't need to write a new regulation to comply with a legal requirement to address air pollution that blows across state lines."[60] EPA administrator Scott Pruitt also decided to suspend the MATS standards—the Mercury Air Toxics rules. The EPA had finalized the MATS in 2011, "requiring coal-burning power plants to reduce emissions of hazardous substances including mercury, lead, arsenic and cadmium by installing control technologies or retiring the plant." Under Pruitt, the EPA decided to delay the effective date; an incredibly confusing round of lawsuits followed. A group of states downwind of major polluters sued the EPA's decision to delay toxic discharge standards for power plants, but a U.S. District Court dismissed that litigation.[61] As of April 2018, the EPA was "still thinking" about the mercury standards, rather than acting on them.

Not just loons and fish are at risk from mercury. Humans suffer as well, for we live in the same watersheds. In one study in Minnesota, nearly 10 percent of the human babies born in the Lake Superior watershed had toxic levels of mercury in their blood at birth, legacies of the fish their mothers ate.[62] As members of the Keweenaw Bay Indian Community ask: "When can we eat the fish?" Fish contamination is an environmental justice issue as well as an ecosystem health issue, and indeed the two are connected.

Acid rain also threatens loons. The same power plants that release mercury also release sulfur and nitrogen oxides that create acid rain. The pollutants fall to earth in snow, rain, and dust particles, eventually washing into lakes and streams across the north. Few if any fish survive in acidic lakes, so loons have less food for their young. Acid rain also increases effective mercury levels in wildlife, because in acidic environments, the less toxic elemental mercury converts more quickly to toxic methylmercury.

Botulism is emerging as another key stressor for loons—and its effects are magnified by climate change. Once largely eradicated from the Great Lakes, botulism returned to Lake Michigan around 2006, carried by invasive zebra mussels and quagga mussels which thrive in warming water. Since 1999, botulism outbreaks have killed more than 100,000 waterbirds in the region. Hardest hit are common loons, white-winged scoters, horned grebes, and long-tailed ducks. In one recent year, 3,000 dead loons were found at a migration stopover site along Lake Michigan; these birds came from a population of about 5,000 total.[63]

Climate Ghosts, Climate Grief

Climate change intensifies the harmful effects of all of these pollutants. As described above, warming lakes experience more active methylation of mercury, and acidic lakes do as well. Climate change is also remobilizing toxics that had been safely sequestered in soil and ice for decades. Higher temperatures encourage the release of these volatile contaminants into the atmosphere, where they can move toward the poles in a process known as global distillation.[64]

Toxics such as DDT, sprayed decades ago, spread as far as the Arctic and Antarctic, where they were sequestered in ice, locked away from doing harm. But melting Antarctic and Arctic ice is remobilizing those DDT residues. Scientist Heidi Geiz's lab showed that DDT levels are increasing in penguins that nest in the Antarctic, even though DDT has not been in widespread use for many years. Those penguins then carry the toxics into broader marine spaces. Similarly, when animals such as fish, loons, and caribou migrate, they likewise become vectors for contamination, bringing chemicals borne in their body fat with them as they swim, fly, or walk.

But migration doesn't just change the environmental distribution of toxic substances; it can also change the effective toxicity of the contaminants for individual animals. An exhausted loon, at the end of its migration, has a suppressed immune system, increasing the hazards of any particular level of toxic substance at a time when an additional day of survival can mean the difference between reproductive success or failure. Additionally, birds fuel their long journeys by burning stored fat—releasing any toxics that had been sequestered in fat back into circulation. When a loon with dwindling energy stores lays her eggs, the toxics in her bloodstream are mobilized into the eggs, exposing her offspring to potentially dangerous levels. A similar process can occur in humans, with mothers passing along 20 percent of our body burden of dioxin, for example, to our offspring.[65]

As long-lived, fish-eating birds at the top of aquatic food chains, loons are the proverbial canaries in the coal mine.[66] Most of the contamination in loon country comes from sources outside the region, often thousands of miles away. Loons are not creatures of remote wildness: they are profoundly entangled in our extractive, fossil-fuel economies.

Bird populations are collapsing extraordinarily quickly. How do we deal with the heartbreak of knowing that so much bird life is vanishing? Rachel Carson, in *Silent Spring*, warned of us a season without the songs of birds—and her passionate call to action averted that dystopia. Yet we always need to keep relearning one lesson of

Silent Spring: the lesson that the health of our waters—and the health of our bodies, our families, our communities—are all interconnected. Loons speak to us of a world on the verge of poisoning, and they ask that we do all we can to slow that toxic wave.

Loons symbolized vision for many northern peoples, and that's still a useful metaphor. Biologist Joe Ricketts argues that loon research is "a way to make the invisible become seen." If we're polluting our lakes to the point where we're killing the loons," he said, "it's a wake up call to us as human beings, that we're causing more damage to our environment than we think we are—and we can't see it except through a bird like the loon."[67]

How do we keep from turning away? How do we look a loon in her eye, see the caribou ghosts, track the lake sturgeon into an uncertain future? How do we comprehend, much less change, the things we have done to the loon, to the north, to our own human selves? The young loons that practice their calls on Seney's lakes bear in their bodies traces of toxics from distant shores. Loons, like other migratory birds, are now part of our climate fingerprint on the world, as they track north, trying to find cooler, cleaner places to continue their ancient migrations.

In "The Gift of the Loon," the loon asks the boy for active engagement. When the loon asks the boy "to guard all the gifts of this world," the boy promises: "I will look after you, always." In addition, the boy pledges, "I shall treasure this world and all I see." Perhaps that is our task now, if we want to stem the tide of rising extinctions. We need to guard all the gifts of this world.

What do we hear, when we listen to the loon? What do we see, when we look at our reflection in its toxic burden, its faltering migrations across paths of poisonous combustion? Will we look after her always? Will we treasure the world and all we see?

NOTES

ONE *Ghosts in the Anthropocene*

1. Steffen, Crutzen, and McNeill, "The Anthropocene."

2. McNeill and Engelke, *The Great Acceleration*.

3. Smith, "Profiles."

4. Palmer, "Profiles."

5. Smith, "Profiles."

6. Wilson, "Mobile Bodies"; Pritchard, "Joining Environmental History with Science and Technology Studies."

7. Latour, *We Have Never Been Modern*. For a discussion of how this affects our toxic understandings, see Nancy Langston, *Toxic Bodies: Hormone Disruptors and the Legacy of DES* (New Haven: Yale University Press, 2010).

8. Rees, "Can Animals Shape Their Own Lives?"

9. Ibid.

10. Gelernter, "Does the Universe Have A Purpose?"

11. Rosenfield, *From Beast-Machine to Man-Machine*. In his book, *Of Wolves and Men*, Barry Lopez discusses how this Cartesian, mechanistic framework influenced Euro-American willingness to slaughter wolves.

12. For a powerful introduction to Indigenous perspectives, see Kimmerer, *Braiding Sweetgrass*.

13. For an early overview of the literature on culture in other animals, see Waal, *The Ape and the Sushi Master*. The field of animal studies has blossomed

in the past two decades, with contributions from philosophy, animal behavior, literature, history, law, and sociology. Useful overviews include: DeMello, *Animals and Society*; McCance, *Critical Animal Studies*; Sorenson, *Critical Animal Studies*; Weil, *Thinking Animals*; and Calarco, *Thinking Through Animals*. I am indebted to Dooren, *Flight Ways*.

14. Cruikshank, "Glaciers and Climate Change," 377. See also Cruikshank, *Do Glaciers Listen?*

15. Grumbine, *Ghost Bears*, 66.

16. Grumbine, *Ghost Bears*; Burnham and Mott, "Timeline."

17. Grumbine, *Ghost Bears*, 67.

18. Ibid.

19. See Langston, *Forest Dreams, Forest Nightmares*; Langston, "Environmental History and Restoration in the Western Forests"; Koontz and Bodine, "Implementing Ecosystem Management in Public Agencies."

20. Langston, *Sustaining Lake Superior*. Grizzlies are still struggling, even though they are no longer ghosts in some regions. They retain threatened status, but the federal government is pushing hard to get them delisted, shorn of protections. Many folks are certain that grizzlies have recovered plenty, enough so that no federal protections are needed and states can move forward with hunts.

Does this pressure to delist the grizzlies mean their populations are approaching the 100,000 estimates of former abundance in North America, the 50,000 in the Lower 48? Not at all. Viable populations don't have to be historic populations, or anything resembling them. The 2019 FWS report estimated that there were about 1,226 grizzlies in the lower 48 (although some of those might have really been resident in Canada).

Not surprisingly, politics, fear of predators, and hopes for industrial development that might bring jobs have complicated grizzly recovery. An increase from 1,000 grizzlies in 1993 to 1,221 in 2019 is at least progress in the right direction. But the populations are still so small they're ghosts in ecological and evolutionary terms. If 50,000 grizzlies shaped these ecosystems—and shaped as well the Indigenous peoples who long figured out ways to coexist with them, in respect, fear, admiration, and fellowship—what does it mean to declare that 1,221 grizzlies are enough?

21. Laundre, Hernandez, and Ripple, "The Landscape of Fear."

22. Jorgensen, "After None"; Francovich, "South Selkirk Mountain Caribou Herd Possibly Extinct."

TWO ○ *Woodland Caribou Histories in the Upper Great Lakes*

1. I discuss this event in more detail in Langston, "Are Woodland Caribou Doomed by Climate Change?"

2. Langston, "Will Woodland Caribou Survive in the Lake Superior Basin?"

3. Rosner, "Pulling Canada's Caribou Back from the Brink."

4. Simkin, "Fish and Wildlife Management Report, April 1, 1960."

5. Environment and Climate Change Canada. "Amended Recovery Strategy for the Woodland Caribou." This document contains useful information about the ecology and habitat requirements of the species. The original recovery strategy can be found at Environment Canada, "Recovery Strategy for the Woodland Caribou (*Rangifer tarandus caribou*), Boreal Population, in Canada."

For an excellent overview of woodland caribou ecology, predation risks, and habitat needs, see Thomas Bergerud's articles over the four decades of his career, including "Antipredator Strategies of Caribou"; "Decline of Caribou in North America Following Settlement"; and "Evolving Perspectives on Caribou Population Dynamics"; and Bergerud et al., "Losing the Predator-prey Space Race Leads to Extirpation of Woodland Caribou from Pukaskwa National Park."

For a critical perspective on habitat and predation risks to woodland caribou in Ontario, see Boan et al., "From Climate to Caribou"; and Boan, Malcolm, and McLaren, "Forest Overstorey and Age as Habitat?"

6. University of Wisconsin-Madison Arboretum, "Caribou Migration."

7. For an excellent popular overview of the technique of using multiple lines of evidence to reconstruct environmental history, see Schulz, "The Earthquake That Will Devastate the Pacific Northwest."

8. See Cleland, *The Prehistoric Animal Ecology and Ethnozoology of the Upper Great Lakes Region*; and Peers, "Ontario Paleo-Indians and Caribou Predation." For an overview of debates about archeological evidence in the eastern woodlands, see Kuehn, "New Evidence for Late Paleoindian-Early Archaic Subsistence Behavior in the Western Great Lakes."

9. Cleland, "Barren Ground Caribou (*Rangifer arcticus*) from an Early Man Site in Southeastern Michigan"; Storck and Spiess, "The Significance of New Faunal Identifications Attributed to an Early Paleoindian (Gainey Complex) Occupation at the Udora Site, Ontario, Canada."

10. Brahic, "Stone Age Hunting Traps Found Deep in Great Lakes." For

research details, see Sonnenburg, Lemke, and O'Shea, *Caribou Hunting in the Upper Great Lakes.*

11. Cruikshank, "Glaciers and Climate Change," *Social Life of Stories,* and *Do Glaciers Listen?* See also Padilla, "Caribou Leadership"; and Padilla and Kofinas, "Letting the Leaders Pass."

12. McInnes, *Sounding Thunder,* chapter 6.

13. Hanson, "The Extraordinary Life of Étienne Brûlé." Étienne Brûlé was sent by Samuel de Champlain from Quebec to traverse the Lake Superior region in 1621. He was probably the first European to traverse the upper Great Lakes country. He left no written reports, and while Champlain wrote about Brûlé's journeys, Champlain himself was focused on minerals and beaver, not other wildlife.

14. Selections from Radisson's journals have been translated and appear in Seno, *Up Country: Voices from the Midwestern Wilderness.* Many records of woodland caribou in the upper Great Lakes exist from the eighteenth century, including those of American trader Jonathan Carver, who in 1766 reported caribou in central Minnesota. Parker, *The Journals of Jonathan Carver and Related Documents, 1766–1770.*

15. Berthel, "Hunting in Minnesota in the Seventies," 262.

16. Breining, *Wild Shore,* 76; pp. 73–81 contain a fascinating account of the author's discussions with Bergerud and others who were doing caribou research on the Slate Islands.

17. Henry, *Alexander Henry's Travels and Adventures in the Years 1760–1776,* 219.

18. Ibid., 219–21.

19. Fashingbauer, "The Woodland Caribou in Minnesota," 142. This essay provides an excellent overview of woodland caribou history in Minnesota.

20. Tanner, *The Falcon.*

21. In the introduction to Tanner's book, Ojibwe author Louise Erdrich provides a powerful analysis of his narrative's cultural context; see especially pp. xi–xiii.

22. In 1848, J. Elliot Cabot wrote about his journey along Lake Superior's Canadian north shore. He noted that "Caribous are found all through this region, but not in great abundance," He added that "An Indian who passed last winter on Isle St. Ignace, killed twenty-five caribou in the course of the winter, and was thought to have done very well." Quoted in Breining, *Wild Shore,* 76.

23. Quoted in Fashingbauer, "The Woodland Caribou in Minnesota," 142.

24. Samuels, *With Rod and Gun in New England and the Maritime Provinces*, 208.

25. Hardy, *Forest Life in Acadie*, 161.

26. Ibid., 163.

27. Ibid., 161.

28. Ibid., 151.

29. Caton, *The Antelope and Deer of America*, 89.

30. Ibid., 90.

31. Whitney, *On Snow-Shoes to the Barren Grounds*, 242.

32. Davis, *Caribou Shooting in Newfoundland*. 155.

33. Samuels, *With Rod and Gun in New England and the Maritime Provinces*, 208.

34. Ibid., 212.

35. Roosevelt, *The Deer Family*, 278. The chapter on caribou was written by Daniel G. Elliot.

36. Bergerud and Mercer, "Caribou Introductions in Eastern North America," 113.

37. Bergerud and Mercer; Manweiler, "Wildlife Management in Minnesota's 'Big Bog,'" and "Minnesota's Woodland Caribou."

38. The classic text is White, *The Middle Ground*. For the Red Lake Nation's history, see Treuer, *Warrior Nation*.

39. Beaulieu, "Mis-qua-Ga-Mi-Wi-Saga-Eh-Ganing History of the Red Lake Band of Chippewa Indians of Minnesota."

40. Ibid.

41. Riis, "Woodland Caribou and Time," 534.

42. Beaulieu, "Mis-qua-Ga-Mi-Wi-Saga-Eh-Ganing History of the Red Lake Band of Chippewa Indians of Minnesota."

43. Ibid.

44. Johnson, "Recollections of the Mammals of Northwestern Minnesota," 450–51.

45. Ibid., 451.

46. Meyer, "The Red Lake Ojibwe," 256.

47. Ibid.

48. Ibid.

49. Ibid.

50. Ibid., 257.

51. Ibid.

52. Bradof, "Ditching of Red Lake Peatland During the Homestead Era," 274, 272; Prince, *Wetlands of the American Midwest*, 281.

53. Averell and McGrew, *The Reaction of Swamp Forests to Drainage in Northern Minnesota*, 45.

54. Prince, *Wetlands of the American Midwest*, 281; Ahrens, "A Contribution to the History of Land Administration in Minnesota."

55. Riis, "Woodland Caribou and Time," 534.

56. Breckenridge, "Trailing the Red Lake Caribou," 10.

57. Ibid.

58. Ibid., 11.

59. Swanson, "The Minnesota Caribou Herd," 416.

60. Ibid., 417.

61. Ibid., 417–418.

62. Ibid., 419.

63. Ibid., 418.

64. Ibid., 419.

65. Manweiler, "Woodland Caribou Study in Northern Minnesota" and "Woodland Caribou from Saskatchewan."

66. Fashingbauer, "The Woodland Caribou in Minnesota," 145.

67. Manweiler, "Woodland Caribou from Saskatchewan," 134.

68. Manweiler, "Woodland Caribou Study in Northern Minnesota," 74.

69. Scott, *Seeing Like a State*.

70. While the Big Bog herd used to migrate, it had abandoned seasonal migration and, by 1938, their range had diminished to "no more than twenty-six miles east and west and from four to nine miles north and south. The winter range is restricted to thirteen miles east and west." Manweiler, "Woodland Caribou Study in Northern Minnesota," 76.

71. Ibid., 77–78.

72. Manweiler, "Woodland Caribou in the Big Bog," 16, 30.

73. Cox, "Woodland Caribou in Minnesota," 138.

74. Ibid., 140.

75. Manweiler, "Minnesota's Woodland Caribou," 35.

76. Fashingbauer, "The Woodland Caribou in Minnesota," 149.

77. Ibid., 163.

78. Ibid., 164.

THREE *Caribou Futures in a Warming World*

1. Langston, "Are Woodland Caribou Doomed by Climate Change?"

2. For useful overviews of the Lake Superior population, see Gogan and Cochrane, "Restoration of Woodland Caribou to the Lake Superior Region." Also helpful is de Vos and Peterson, "A Review of the Status of Woodland Caribou (*Rangifer caribo*u) in Ontario"; and Bergerud et al., "Woodland Caribou Persistence and Extirpation in Relic Populations on Lake Superior." For broader Canadian context, see Badiou et al., "Keeping Woodland Caribou in the Boreal Forest." Discussion of attempts to put into place the federal Canadian recovery plan can be found at Armstrong et al., "Caribou Conservation and Recovery in Ontario"; and Environment Canada, "Recovery Strategy for the Woodland Caribou (*Rangifer tarandus caribou*), Boreal Population, in Canada." For an excellent overview of caribou translocations, see Eason, "Caribou Introductions."

3. Bergerud et al., "Woodland Caribou Persistence and Extirpation in Relic Populations on Lake Superior."

4. Gogan and Cochrane, "Restoration of Woodland Caribou to the Lake Superior Region."

5. Breining, *Wild Shore*, 46–50.

6. See Conway and Conway, *Spirits on Stone*. Henry Schoolcraft visited these pictographs and wrote the first Euro-American records of their patterns and significance for the Anishinaabeg in *Historical and Statistical Information Respecting the History, Condition and Prospects of the Indian Tribes of the United States*.

7. Boan et al., "From Climate to Caribou."

8. Environment Canada, "Recovery Strategy for the Woodland Caribou (*Rangifer tarandus caribou*), Boreal Population, in Canada."

9. Masood et al., "An Uncertain Future for Woodland Caribou (*Rangifer tarandus caribou*)."

10. Lopez, *Of Wolves and Men*; Coleman, *Vicious*.

11. Leopold, *A Sand County Almanac and Sketches Here and There*, 130. Leopold was also fascinated by woodland caribou, particularly on Isle Royale. The classic analysis is Flader, *Thinking Like a Mountain*. The Aldo Leopold Foundation produced an interesting film analyzing Leopold's moment of insight into wolves: Steinke, *Green Fire: Aldo Leopold and a Land Ethic for Our Time*.

12. See Vucetich and Nelson, "Wolf Hunting and the Ethics of Predator Control." In *Natural*, ethics and religion scholar Alan Levinovitz argues that a useful ethic might be one that's place-based, local, admits to uncertainty and complexity, and takes actions that are modest and reversible.

13. Debates over the multiple causes of woodland caribou decline are prolific in the scientific literature. Thomas Bergerud long argued that predation played the key role; see, for example, "Decline of Caribou in North America Following Settlement," "Evolving Perspectives on Caribou Population Dynamics," and "The Caribou Conservation Conundrum." Another important perspective stresses the roles of habitat disturbance; one of the first scientists to make this case was Cringan, "History, Food Habits and Range Requirements of the Woodland Caribou of Continental North America." Most biologists agree with Bergerud that, in fragmented landscapes, wolf density probably needs to be less than 6.5 wolves per 1000 square kilometers for caribou to persist.

14. Weber, "Increase Wolf Cull, Pen Pregnant Cows to Save Endangered Caribou."

15. The controversial study was Serrouya et al., "Saving Endangered Species Using Adaptive Management." Serrouya et al., "Predicting the Effects of Restoring Linear Features on Woodland Caribou Populations," 6181.

16. Harding et al., "No Statistical Support for Wolf Control and Maternal Penning as Conservation Measures for Endangered Mountain Caribou."

17. Imbler, "How a Simple Statistical Error Killed 463 Wolves."

18. Bird, "Killing Wolves Won't Save Caribou Herds, Experts Say."

19. Page, "Efforts to Save Woodland Caribou in Northern Quebec Too Costly, Says Province | CBC News."

20. Droitsch, "Canada Chooses Tar Sands over Caribou."

21. Boutin et al., "Why Are Caribou Declining in the Oil Sands."

22. Imbler, "How a Simple Statistical Error Killed 463 Wolves" (citing Fletcher, "B.C. Interior Caribou Protection Area Big Enough, Minister Says").

23. Adapted from Langston, "Paradise Lost."

24. Huff and Thomas, "Lake Superior Climate Change Impacts and Adaptation."

25. Ibid.

26. Thorpe et al., *Ecological and Policy Implications of Introducing Exotic Trees for Adaptation to Climate Change in the Western Boreal Forest.*

27. Saskatchewan Environmental Society, "Climate Change and Saskatchewan's Boreal Forest," 2.

28. Nickens, "Paper Chase."

29. Langston, *Sustaining Lake Superior*, and Huff and Thomas, "Lake Superior Climate Change Impacts and Adaptation."

30. Grayson and Delpech, "Pleistocene Reindeer and Global Warming."

31. Langston, "Are Woodland Caribou Doomed by Climate Change?"

32. Geist, "Of Reindeer and Man, Modern and Neanderthal," 52.

33. Vitebsky, *The Reindeer People*.

34. Nelson, "Eskimo Science," 91.

35. Nelson, quoted in Lewis, "Deer to Our Culture."

36. Berger, *Why Look at Animals*, 4–8.

37. Bennett, "How Animals Made Us Human."

38. Ibid. For an overview of Shipman's excellent work, see Shipman, *The Animal Connection*.

39. For some of the fascinating recent material rethinking the history of domestication, see Budiansky, *The Covenant of the Wild*; Hobgood-Oster, *A Dog's History of the World*; Lescureux, "Beyond Wild and Domestic"; Losey et al., "Domestication and the Embodied Human–Dog Relationship"; and Dugatkin and Trut, *How to Tame a Fox (and Build a Dog)*.

40. Westropp, "On the Sequence of the Phases of Civilisation, and Contemporaneous Implements," cxcii.

41. Foster, "Recent Advances in Geology," 464.

42. See Piper and Sandlos, "A Broken Frontier"; Demuth, *Floating Coast*; Wills, "The Reindeer Games"; and Willis, "A New Game in the North."

43. Baird, *Report of the Secretary of Agriculture*, 108.

44. Shields, *The Big Game of North America*, 16.

45. Rudolf, *History of the Lake States Forest Experiment Station*. According to Rudolf, "An interesting study in 1932 involved a herd of reindeer introduced into the Superior National Forest in 1930. Many of the animals were lost and the Station study indicated that the project was not feasible because of limited food supply." Note: I could not go to the station's archives and track that study down because of COVID-19 restrictions. But I did consult two retired biologists who had long worked on caribou in the forest, and they hadn't heard of a reindeer introduction.

46. Vitebsky, *The Reindeer People*.

47. Langston, "Mining the Boreal North."

48. Gwich'in Steering Committee, "A Moral Choice for the United States," 4.

49. Wray and Parlee, "Ways We Respect Caribou," 68.

50. Solnoi, Tsogtsalkhan, and Plumley, "Following the White Stag."

51. See Bruno, *The Nature of Soviet Power*; Demuth, *Floating Coast*; and Vitebsky, *The Reindeer People*.

52. Kingsley and Levene, "Nomads No More."

53. Spence, *Dispossessing the Wilderness*. And also see Jacoby, *Crimes Against Nature*.

54. Te Beest et al., "Reindeer Grazing Increases Summer Albedo by Reducing Shrub Abundance in Arctic Tundra."

55. This section is adapted from my earlier essay, Langston, "Mining the Boreal North."

56. Anderson and Hausfather, "Confronting Limits to Climate Adaptation in the Natural World."

FOUR *Indigenous Communities and Lake Sturgeon Restoration*

1. Carey, *The Philosopher Fish*, 5.

2. Waldman, "The Lofty Mystery of Why Sturgeon Leap."

3. Kallok, "A Whopper of a Recovery;" and Heasley, *The Accidental Reef*, chapter 6.

4. See Auer and Dempsey, *The Great Lake Sturgeon*; and Kline, *People of the Sturgeon*.

5. Carey, *The Philosopher Fish*, 3.

6. Gardiner, "Sturgeons as Living Fossils," 148.

7. Fuller, "Lake Sturgeon (*Acipenser fulvescens*)—Species Profile."

8. Packer, "Manitoba History: Glacial Lake Agassiz."

9. Kallok, "A Whopper of a Recovery."

10. Carey, *The Philosopher Fish*, 6.

11. Waldman, "The Lofty Mystery of Why Sturgeon Leap," and Catesby, *The Natural History of Carolina, Florida and the Bahama Islands*, vol. 1 (1729–1732): xxxiii.

12. Sulak et al., "Why Do Sturgeons Jump?" and Waldman, "The Lofty Mystery of Why Sturgeon Leap."

13. Auer and Baker, "Duration and Drift of Larval Lake Sturgeon in the Sturgeon River, Michigan."

14. Holtgren, Ogren, and Whyte, "Renewing Relatives." I rely heavily on this article's powerful analysis. Other core works in Anishinaabeg material use

and relations with sturgeon include Holzkamm, "Sturgeon Utilization by the Rainy River Ojibwa Bands"; Holzkamm and McCarthy, "Potential Fishery for Lake Sturgeon (*Acipenser fulvescens*) as Indicated by the Returns of the Hudson's Bay Company Lac La Pluie District"; Holzkamm and Waisberg, "Native American Utilization of Sturgeon"; Holzkamm, Lytwyn, and Waisberg, "Rainy River Sturgeon"; and Hannibal-Paci, "His Knowledge and My Knowledge" and "Historical Representations of Lake Sturgeon by Native and Non-Native Artists."

15. Spratt, "The Return of Namé 'King of Fish' to the Red River Basin | U.S. Fish and Wildlife Service Midwest Region," June 5, 2017.

16. Loew, *Indian Nations of Wisconsin*, 60.

17. Holzkamm and Waisberg, "Native American Utilization of Sturgeon."

18. Radisson, translated in Joseph, *Up Country*, 9.

19. Allouez, translated in Joseph, *Up Country*, 32.

20. "Lake Sturgeon Rehabilitation."

21. "A Brief History of Lake Sturgeon in Menonminee [sic] and Ojibwe Culture."

22. Translated in Joseph, *Up Country*, 42.

23. "Lake Sturgeon Restoration."

24. Spratt, "The Return of Namé 'King of Fish' to the Red River Basin."

25. Holtgren, Ogren, and Whyte, "Renewing Relatives."

26. Ibid.

27. Tanner, *The Falcon*, 25.

28. Ibid., 39.

29. Ibid., 61.

30. Ibid., 62.

31. Tanner, 252.

32. "A Brief History of Lake Sturgeon in Menonminee [sic] and Ojibwe Culture" and Kallok, "A Whopper of a Recovery."

33. "A Brief History of Lake Sturgeon in Menonminee [sic] and Ojibwe Culture."

34. Milner, "U.S. Fish Commission: Report of the Commissioner for 1872–73," 72.

35. Ibid., 74.

36. Williams, "Recovery."

37. Kallok, "A Whopper of a Recovery."

38. Williams, "Recovery."

39. Milner noted that "As an article of food they are not generally popular. But few people in the cities know the modes of cooking that make their meat a palatable dish…. the bouillon, when carefully prepared by skimming off the oil, is very much like chicken-soup. A very good pickled meat is made of it by boiling it and preserving it in vinegar. But the best form of preparing sturgeon is by smoking it." Milner, "U.S. Fish Commission: Report of the Commissioner for 1872–73," 71.

40. Kallok, "A Whopper of a Recovery."

41. Holzkamm, "Sturgeon Utilization by the Rainy River Ojibwa Bands."

42. Milner, "U.S. Fish Commission: Report of the Commissioner for 1872–73," 75. The whitefish fishermen were thrilled at this decline, Milner noted.

43. Ibid., 23.

44. See Langston, *Sustaining Lake Superior*.

45. Kallok, "A Whopper of a Recovery."

46. Langston, *Sustaining Lake Superior*.

47. Creger, "Taming Water, A Diverting Story of Ebbs & Flows."

48. "Plan to Solve Great Lakes Problem." For much more on proposals to divert water from the Great Lakes, see Annin, *The Great Lakes Water Wars*.

49. Williams, "Recovery."

50. Kallok, "A Whopper of a Recovery."

51. The classic analyses are Taylor, *Making Salmon*; McEvoy, *The Fisherman's Problem*; and White, *The Organic Machine*.

52. Momaday, quoted in Burns, *The West*, "Episode 5 (1868–1874)."

53. White Horse quoted in Johnsgard, *The Niobrara*, 116.

54. White, *The Organic Machine*, 47–48.

55. Bottom, "Restoring Salmon Ecosystems," 170.

56. Duncan, "Second Coming," 202.

57. Loew, *Indian Nations of Wisconsin*, 66–68.

58. Loew and Thannum, "After the Storm."

59. Holtgren, Ogren, and Whyte, "Renewing Relatives." This section is heavily indebted to their analysis.

60. Ibid.

61. Ibid.

62. Ibid.

63. Ibid.

64. Ibid.

65. Spratt, "The Return of Namé 'King of Fish' to the Red River Basin."

66. Ibid.

67. Kallok, "A Whopper of a Recovery."

68. "A Brief History of Lake Sturgeon in Menonminee [sic] and Ojibwe Culture." See also Frechette and Hoffman, "The Menominee Clans Story."

69. Holzkamm and Waisberg, "Native American Utilization of Sturgeon."

70. "A Brief History of Lake Sturgeon in Menonminee [sic] and Ojibwe Culture."

71. Bruch and Koenigs, "Welcoming Back Namao," 18.

72. Ibid., 17.

73. Ibid., 18–19.

74. Ibid., 19.

75. Castro, Hossain, and Tytelman, "Arctic Ontologies," 100.

76. Langston and Christen, "Conservation Policies Threaten Indigenous Reindeer Herders in Mongolia."

77. Westley et al., "Collective Movement in Ecology."

78. Holtgren, Ogren, and Whyte, "Renewing Relatives."

79. "Resilience of the Lake Sturgeon."

80. Crawley, "Bringing 'Two Eyed Seeing'—Indigenous Knowledge and Science—to Fisheries Conservation." For more detailed discussions, see Hassell, *Two-Eyed Seeing*; McMillan and Prosper, "Remobilizing netukulimk"; Rayne et al., "Centring Indigenous Knowledge Systems to Re-imagine Conservation Translocations"; and Reid, et al., "Two-eyed Seeing."

FIVE The Gift of the Loon

1. Langston, "Environmental History and Restoration in the Western Forests." Critiques of the wilderness idea have flourished since the publication of William Cronon, "The Trouble with Wilderness." But as problematic as wilderness ideas are, the connection between loons and wildness persists. See Klein, "Loonacy," for an excellent overview. For a lovely evocation of loons and wildness, see Olson, *Singing Wilderness*.

2. Guynup, "Loons Sound Alarm on Mercury Contamination."

3. Siewert, "In Wisconsin, Hate Makes a Comeback." In 2020, Klan activities have increased in Mercer and across northern Wisconsin. Michael McQueeney, owner of Antler's Pub in Mercer, has been a grand dragon with the Klan and continues to organize white supremacist rallies in the area.

4. "The Common Loon."

5. Evers, *Conserve the Call.*

6. Bjorhus, "Loons Likely to Disappear from Minnesota, New Report Warns."

7. "Survival by Degrees: 389 Bird Species on the Brink."

8. Pennisi, "Three Billion North American Birds Have Vanished since 1970, Surveys Show."

9. Gross, "Feeding Wild Birds Can Carry Risks."

10. For a history of avian extinction concerns, see Lewis, *Belonging on an Island.*

11. "The Common Loon."

12. Nelson, *Make Prayers to the Raven,* 16.

13. There are many versions of this story. This one is from Friedman and Johnson, "The Gift of the Loon: An Inuit Legend." A particularly powerful version can be viewed in the film and book by Alethea Arnaquq-Baril, *The Blind Boy and the Loon.*

14. Auger, *The Way of Inuit Art,* 45.

15. Nelson, *Make Prayers to the Raven,* 85.

16. Cannon, "Alaska Athabascan Stellar Astronomy," 9; citing McClellan, *My Old People Say.*

17. King, Pauksztat, and Storrie, *Arctic Clothing,* 66.

18. Nelson, *Make Prayers to the Raven,* 86.

19. King, Pauksztat, and Storrie, *Arctic Clothing,* 66.

20. Auger, *The Way of Inuit Art,* 45.

21. Nelson, *Make Prayers to the Raven,* 87.

22. Klein, "Loonacy."

23. Nootchtai, M. "Why the Loon Can't Walk." Thanks to K. Brosemer for this story.

24. Klein, "Loonacy."

25. Harland-Haughey, "The Broken Bird."

26. Lopez, *Of Wolves and Men,* 210.

27. Thoreau, *Walden,* 152.

28. Ibid.

29. Facebook post written by Damon McCormick, of Common Coast Research & Conservation; www.facebook.com/seneyrefuge/photos/a.136852396351905/2967087446661705.

30. U.S. Government, "Treaty with the Ottawa, Etc." The Bay Mills Indian Community, Sault Ste. Marie Tribe of the Chippewa Indians, Grand Traverse Band of Ottawa and Chippewa Indians, Little River Band Ottawa Indians, and Little Traverse Bay Band of Odawa Indians continue to exercise hunting

and fishing rights on lands within the 1836 treaty. In November 2007, the United States, State of Michigan, and the five tribes signed an Inland Consent Decree, which affirms tribal rights within the refuge. See Casselman, "Seney National Wildlife Refuge Comprehensive Conservation Plan," 54.

31. "Seney National Wildlife Refuge Annual Narrative Report."

32. Ibid.

33. Ibid. For a detailed account of wildlife refuge wetland management, see Langston, *Where Land and Water Meet*. Native Americans were allowed to join the CCC starting in 1933, and many served at Camp Marquette in the Upper Peninsula, where they replanted trees, improved timber stands, and constructed roads. One tribal member of this camp allegedly remarked, "The white man stole our land in the first place, cut off the timber, and now they are making us plant it again." Rosentreter, "Roosevelt's Tree Army."

For a local history of CCC camps in the region, see Chabot, *Saving Our Sons*. For a scholarly history of the CCC and its role in conservation, see Maher, *Nature's New Deal*.

34. McCormick, Kaplan, and Tischler, "Common Loon Research at Seney National Wildlife Refuge."

35. Quoted in Klein, "Loonacy."

36. U.S. Geological Survey, Upper Midwest Environmental Sciences Center. "Loon Study—Frequently Asked Questions."

37. Rulseh, "Happy Anniversary, LoonWatch."

38. McCormick, Kaplan, and Tischler, "Common Loon Research at Seney National Wildlife Refuge," 5–6.

39. Ibid., 12.

40. Ibid., 4.

41. Barcott, "The Lure of the Common Loon."

42. U.S. Geological Survey, Upper Midwest Environmental Sciences Center, "Common Loon Migration Study—Update."

43. Murray, *Adventures in the Wilderness or Camp-Life in the Adirondacks*, 106-113.

44. This phrase has actually been trademarked by one fishing lure company: Esca Global, http://escaglobal.eu/.

45. Dorsey, The Dawn of Conservation Diplomacy, 215.

46. Olson, Loftin, and Goodwin, "Biological, Geographical, and Cultural Origins of the Loon Hunting Tradition in Carteret County, North Carolina."

47. The quote is from an unnamed observer mentioned in Eastman, *The Eastman Guide to Birds*, 219. For details on the spray campaigns, see Langston, *Sustaining Lake Superior*, 54–58.

48. Klein, "Loonacy."

49. Evers, *Conserve the Call.*

50. Evers, *Conserve the Call* and Evers, et al., "Geographic Trend in Mercury Measured in Common Loon Feathers and Blood."

51. Northland College, "Protect Loons."

52. Mele and Mele, *Polluting for Pleasure*, 20. Thanks to K. Brosemer for this detail.

53. McCormick, Kaplan, and Tischler, "Common Loon Research at Seney National Wildlife Refuge," 5. For details of the study, see McCormick, Kaplan, and Tischler, "Mercury Exposure in Common Loons at Seney National Wildlife Refuge."

54. Pokras is cited in Guynup, "Loons Sound Alarm on Mercury Contamination." For original research on mercury in loons, see Evers, et al., "Geographic Trend in Mercury Measured in Common Loon Feathers and Blood"; Bhavsar, et al., "Changes in Mercury Levels in Great Lakes Fish Between 1970s and 2007"; Best, "Assessment of Mercury in Edible Fish Fillets at Seney National Wildlife Refuge"; McCormick, Kaplan, and Tischler, "Mercury Exposure in Common Loons at Seney NWR."

55. Tozer et al., "Common Loon Reproductive Success in Canada."

56. Barcott, "The Lure of the Common Loon."

57. Klein, "Loonacy."

58. Tozer et al., "Common Loon Reproductive Success in Canada."

59. Sheehan, "Acid Rain, Mercury and Loons . . . Oh My!"

60. Cama, "EPA Says It Doesn't Need New 'Good Neighbor' Air Pollution Rule."

61. Reilly, "EPA 'Still Thinking About' Obama Mercury Standards."

62. McCann, "Mercury Levels in Blood from Newborns in Lake Superior Basin." Thanks to Dr. V. Gagnon for conversations on tribal fish consumption.

63. Barcott, "The Lure of the Common Loon." For statistics on botulism, see Michigan Sea Grant, "Avian Botulism."

64. See Langston, *Sustaining Lake Superior.* Thanks to K. Brosemer for this information.

65. Yaktine, Harrison, and Lawrence, "Reducing Exposure to Dioxins and Related Compounds through Foods in the Next Generation," 403–409.

66. Barcott, "The Lure of the Common Loon."

67. Ibid.

BIBLIOGRAPHY

Ahrens, Marjorie Henkel. "A Contribution to the History of Land Administration in Minnesota: The Origins of the Red Lake Game Preserve." M.A. thesis, University of Minnesota, 1987.

Anderson, Lauren, and Zeke Hausfather. "Confronting Limits to Climate Adaptation in the Natural World." The Breakthrough Institute, May 27, 2020. https://thebreakthrough.org/issues/conservation/confronting-adaptation-limits.

Annin, Peter. *The Great Lakes Water Wars*. Washington, DC: Island Press, 2009.

Armstrong, Ted (E. R.), Michael Gluck, Glen Hooper, Iain Mettam, Gerald D. Racey, and Marc Rondeau. "Caribou Conservation and Recovery in Ontario: Development and Implementation of the Caribou Conservation Plan." *Rangifer* 32, no. 2 (March 8, 2012): 145–57. https://doi.org/10.7557/2.32.2.2266.

Arnaquq-Baril, Alethea. *The Blind Boy and the Loon*. Toronto: Inhabit Media, 2014.

Auer, N. A., and E. A. Baker. "Duration and Drift of Larval Lake Sturgeon in the Sturgeon River, Michigan." *Journal of Applied Ichthyology* 18, nos. 4–6 (2002): 557–64. https://doi.org/10.1046/j.1439-0426.2002.00393.x.

Auer, Nancy, and Dave Dempsey. *The Great Lake Sturgeon*. East Lansing: Michigan State University Press, 2013.

Auger, Emily Elisabeth. *The Way of Inuit Art: Aesthetics and History in and Beyond the Arctic.* Jefferson, NC: McFarland, 2005.

Averell, James Lawrence, and Paul Cornelius McGrew. *The Reaction of Swamp Forests to Drainage in Northern Minnesota.* St. Paul, MN: Department of Drainage and Waters, 1929.

Badiou, Pascal, et al. "Keeping Woodland Caribou in the Boreal Forest: Big Challenge, Immense Opportunity." International Boreal Conservation Panel, July 2011. http://www.borealscience.org/wp-content/uploads /2012/06/brief-woodlandcaribou.pdf.

Baird, Spencer. *Report of the Secretary of Agriculture.* U.S. Government Printing Office, 1862.

Barcott, Bruce. "The Lure of the Common Loon." *Audubon,* October 2, 2013. https://www.audubon.org/news/the-lure-common-loon.

Beaulieu, Kathryn. "Mis-qua-Ga-Mi-Wi-Saga-Eh-Ganing History of the Red Lake Band of Chippewa Indians of Minnesota." *Red Lake History Project.* Accessed July 22, 2020. http://www.rlnn.org/MajorSponsors /HistoryProjectBeginning.html.

Bennett, Drake. "How Animals Made Us Human." *Boston Globe,* September 12, 2010. http://archive.boston.com/bostonglobe/ideas/articles/2010/09/12 /what_explains_the_ascendance_of_homo_sapiens_start_by_looking_at _our_pets/?page=3.

Berger, John. *Why Look at Animals?* London: Penguin Books, 2009.

Bergerud, Arthur T. "Antipredator Strategies of Caribou: Dispersion along Shorelines." *Canadian Journal of Zoology* 63, no. 6 (1985): 1324–29.

———. "Decline of Caribou in North America Following Settlement." *The Journal of Wildlife Management* 38, no. 4 (1974): 757–70.

———. "Evolving Perspectives on Caribou Population Dynamics, Have We Got It Right Yet?" *Rangifer* 16, no. 4 (January 1, 1996): 95–116. https://doi .org/10.7557/2.16.4.1225.

———. "The Caribou Conservation Conundrum." In *The Real Wolf: The Science, Politics, and Economics of Coexisting with Wolves in Modern Times,* edited by Ted B. Lyon and Will N. Graves. New York: Skyhorse Publishing, 2018.

Bergerud, A. T., W. J. Dalton, H. Butler, L. Camps, and R. Ferguson. "Woodland Caribou Persistence and Extirpation in Relic Populations on Lake Superior." *Rangifer* 17 (2007): 57–78. http://dx.doi.org/10.7557 /2.27.4.321.

Bergerud, A. T., Brian E. McLaren, Ludvik Krysl, Keith Wade, and William Wyett. "Losing the Predator-prey Space Race Leads to Extirpation of Woodland Caribou from Pukaskwa National Park." *Écoscience* 21, no. 3–4 (September 2014): 374–86. https://doi.org/10.2980/21-(3-4)-3700.

Bergerud, A. T., and W. E. Mercer. "Caribou Introductions in Eastern North America." *Wildlife Society Bulletin* 17, no. 2 (1989): 111–20.

Berthel, Mary Wheelhouse. "Hunting in Minnesota in the Seventies." *Minnesota History* 16, no. 3 (1935): 259–71.

Best, David A. "Assessment of Mercury in Edible Fish Fillets at Seney National Wildlife Refuge." Environmental Contaminants Program, Division of Ecological Services, East Lansing Field Office, U.S. Fish and Wildlife Service, September 1999. https://ecos.fws.gov/ServCat /DownloadFile/21498

Bhavsar, Satyendra P., Sarah B. Gewurtz, Daryl J. McGoldrick, Michael J. Keir, and Sean M. Backus. "Changes in Mercury Levels in Great Lakes Fish Between 1970s and 2007." *Environmental Science and Technolology* 44, no. 9 (2010): 3273–79. https://doi.org/10.1021/es903874x.

Bird, Hilary. "Killing Wolves Won't Save Caribou Herds, Experts Say." *CBC News*, February 27, 2019. https://www.cbc.ca/news/canada/north/wolf-cull -misses-mark-experts-say-1.5030249.

Bjorhus, Jennifer. "Loons Likely to Disappear from Minnesota Due to Climate Change, New Report Warns." *Star Tribune*, October 12, 2019; https:// www.startribune.com/loons-likely-to-disappear-from-minnesota-due-to -climate-change-new-report-warns/562874132/.

Boan, Julee J., Jay R. Malcolm, and Brian E. McLaren. "Forest Overstorey and Age as Habitat? Detecting the Indirect and Direct Effects of Predators in Defining Habitat in a Harvested Boreal Landscape." *Forest Ecology and Management* 326 (August 2014): 101–8. https://doi.org/10.1016/j.foreco .2014.03.052.

Boan, Julee J., Jay R. Malcolm, Mallory D. Vanier, Dave L. Euler, and Faisal M. Moola. "From Climate to Caribou: How Manufactured Uncertainty Is Affecting Wildlife Management." *Wildlife Society Bulletin* 42, no. 2 (June 2018): 366–81. https://doi.org/10.1002/wsb.891.

Bottom, Daniel L. "Restoring Salmon Ecosystems: Myth and Reality." *Restoration & Management Notes* 13, no. 2 (1995): 162–70.

Boutin, Stan, Mark Boyce, Mark Hebblewhite, Dave Hervieux, Kyle Knopff, Maria Latham, Andrew Latham, John Nagy, Dale Seip, and Robert

Serrouya. "Why Are Caribou Declining in the Oil Sands?" *Frontiers in Ecology and the Environment* 10 (March 1, 2012): 65–67. https://doi.org /10.2307/41480001.

Bradof, Kristine L. "Ditching of Red Lake Peatland During the Homestead Era." In *Patterned Peatlands of Minnesota*, edited by H. E. Wright, Barbara Coffin, and Norman Aaseng, 263–84. Minneapolis: University of Minnesota Press, 1992.

Brahic, Catherine. "Stone Age Hunting Traps Found Deep in Great Lakes." *New Scientist*, June 9, 2009. https://www.newscientist.com/article/dn17275 -stone-age-hunting-traps-found-deep-in-great-lakes/.

Breckenridge, W. J. "Trailing the Red Lake Caribou." *The Minnesota Conservationist* 26 (1935): 10–11.

Breining, Greg. *Wild Shore: Exploring Lake Superior by Kayak*. Minneapolis: University of Minnesota Press, 2000.

"A Brief History of Lake Sturgeon in Menonminee [sic] and Ojibwe Culture." *World History* (blog), May 1, 2018. https://worldhistory.us/american-history /native-american-history/a-brief-history-of-lake-sturgeon-in-menonminee -and-ojibwe-culture.php.

Bruch, Ron, and Ryan Koenigs. "Welcoming Back Namao." *Wisconsin Natural Resources Magazine*, August 2013, 17–19. http://www.onwisconsinoutdoors .com/InlandFishing/sturgeonReturnsKeshenaFalls.

Bruno, Andy. *The Nature of Soviet Power: An Arctic Environmental History*. New York: Cambridge University Press, 2016.

Budiansky, Stephen. *The Covenant of the Wild: Why Animals Chose Domestication*. London: Weidenfeld & Nicholson, 1994.

Burnham, Josh, and Nick Mott. "Timeline: A History of Grizzly Bear Recovery in the Lower 48 States." *Montana Public Radio Here and Now*, May 14, 2019. https://www.mtpr.org/post/timeline-history-grizzly-bear -recovery-lower-48-states.

Burns, Ken. *The West*, "Episode 5 (1868–1874), 2001. http://www.shoppbs.pbs .org/weta/thewest/program/episodes/five/woundinheart.htm.

Calarco, Matthew. *Thinking Through Animals: Identity, Difference, Indistinction*. Redwood City, CA: Stanford Briefs, 2015.

Cama, Timothy. "EPA Says It Doesn't Need New 'Good Neighbor' Air Pollution Rule." *The Hill*, June 29, 2018. https://thehill.com/policy/energy -environment/394860-epa-says-it-I-need-new-good-neighbor-air -pollution-rule.

Cannon, Christopher. "Alaska Athabascan Stellar Astronomy." Master's thesis, University of Alaska Fairbanks, 2014.

Carey, Richard Adams. *The Philosopher Fish: Sturgeon, Caviar, and the Geography of Desire*. Berkeley, CA: Counterpoint Press, 2005.

Carson, Rachel. *Silent Spring*. New York: Houghton Mifflin Harcourt, 1962.

Casselman, Tracy. "Seney National Wildlife Refuge Comprehensive Conservation Plan," U.S. Fish and Wildlife Service, 2009.

Castro, Damian, Kamrul Hossain, and Carolina Tytelman. "Arctic Ontologies: Reframing the Relationship between Humans and Rangifer." *Polar Geography* 39, no. 2 (2016): 98–112.

Catesby, Mark. *The Natural History of Carolina, Florida and the Bahama Islands*. Vol. 1 (1729–1732). London: Royal Society, 1729. https://www.biodiversity library.org/item/126524.

Caton, John Dean. *The Antelope and Deer of America*. New York: Hurd and Houghton, 1877. http://archive.org/details/antelopeanddeer00catogoog.

Chabot, Larry. *Saving Our Sons: How the Civilian Conservation Corps Rescued a Generation of Upper Michigan Men*. Marquette, MI: North Harbor Publishing, 2009.

Cleland, Charles E. "Barren Ground Caribou (*Rangifer arcticus*) from an Early Man Site in Southeastern Michigan." *American Antiquity* 20, no. 3 (January 1965): 350–51. https://doi.org/10.2307/278816.

———. *The Prehistoric Animal Ecology and Ethnozoology of the Upper Great Lakes Region*. Ann Arbor: University of Michigan Press, 1966.

Coleman, Jon T. *Vicious: Wolves and Men in America*. New Haven: Yale University Press, 2004.

"The Common Loon." National Park Service, April 14, 2015. https://www.nps .gov/gaar/learn/nature/common-loon-1.htm.

Conway, Thor, and Julie Conway. *Spirits on Stone: The Agawa Pictographs*. San Luis Obispo, CA: Heritage Discoveries, 1991.

Cox, W. T. "Woodland Caribou in Minnesota." *Soil Conservation* 6 (1939): 138–43, 156.

Crawley, Mark. "Bringing 'Two Eyed Seeing'—Indigenous Knowledge and Science—to Fisheries Conservation." *CBC Radio*, February 14, 2020. https://www.cbc.ca/radio/quirks/feb-15-agriculture-moving-north -arrokoth-s-secrets-the-microbiome-for-flight-and-more-1.5463847 /bringing-two-eyed-seeing-indigenous-knowledge-and-science-to -fisheries-conservation-1.5463853.

Creger, Michael. "Taming Water, A Diverting Story of Ebbs & Flows." *Lake Superior Magazine*, October 15, 2018. https://www.lakesuperior.com/api /content/7a8f7488-8e8f-11e8-b207-12408cbff2b0/.

Cringan, Alexander Thorn. "History, Food Habits and Range Requirements of the Woodland Caribou of Continental North America." *Transactions of the North American Wildlife Conference* 22 (1957): 485–501.

Cronon, William. "The Trouble with Wilderness: Or, Getting Back to the Wrong Nature." *Environmental History* 1, no. 1 (1996): 7–28.

Cruikshank, Julie. *Do Glaciers Listen? Local Knowledge, Colonial Encounters, and Social Imagination*. Vancouver: University of British Columbia Press, 2005.

———. "Glaciers and Climate Change: Perspectives from Oral Tradition." *Arctic* 54, no. 4 (2001): 377–93.

———. *Social Life of Stories: Narrative and Knowledge in the Yukon Territory*. Vancouver: University of British Columbia Press, 2000.

Davis, Samuel T. *Caribou Shooting in Newfoundland: With a History of England's Oldest Colony from 1001 to 1895*. Lancaster, PA: The New Era Printing House, 1895. http://www.biodiversitylibrary.org/item/68533.

DeMello, Margo. *Animals and Society: An Introduction to Human-Animal Studies*. New York: Columbia University Press, 2012.

Demuth, Bathsheba. *Floating Coast: An Environmental History of the Bering Strait*. New York: Norton, 2019.

de Vos, A., and Randolph L. Peterson. "A Review of the Status of Woodland Caribou (*Rangifer caribou*) in Ontario." *Journal of Mammalogy* 32, no. 3 (1951): 329–37.

Dooren, Thom van. *Flight Ways: Life and Loss at the Edge of Extinction*. New York: Columbia University Press, 2014.

Dorsey, Kurkpatrick. *The Dawn of Conservation Diplomacy: U.S.-Canadian Wildlife Protection Treaties in the Progressive Era*, Seattle: University of Washington Press, 2010.

Droitsch, Danielle. "Canada Chooses Tar Sands over Caribou." *NRDC* (blog), August 29, 2011. https://www.nrdc.org/experts/danielle-droitsch/canada -chooses-tar-sands-over-caribou.

Dugatkin, Lee Alan, and Lyudmila Trut. *How to Tame a Fox (and Build a Dog): Visionary Scientists and a Siberian Tale of Jump-Started Evolution*. Chicago: University of Chicago Press, 2017.

Duncan, David James. "Second Coming." *Sierra Magazine*, April 2000. https:// vault.sierraclub.org/sierra/200003/salmon1.asp.

Eason, Gordon. "Caribou Introductions—Wawa District: Implications for
    Caribou Restoration in the Lake Superior Area." Report, Ontario Ministry
    of Natural Resources, Wawa, Ontario, 2012. Copy in the possession of the
    author.
Eastman, John. *The Eastman Guide to Birds: Natural History Accounts for 150
    North American Species.* Mechanicsburg, PA: Stackpole Books, 2000.
Environment and Climate Change Canada. "Amended Recovery Strategy for
    the Woodland Caribou (*Rangifer tarandus caribou*), Boreal Population, in
    Canada [Proposed] 2019." Species at Risk Act: Recovery Strategy Series,
    July 9, 2019. https://www.canada.ca/en/environment-climate-change
    /services/species-risk-public-registry/recovery-strategies/woodland-caribou
    -boreal-2019.html
Environment Canada. "Recovery Strategy for the Woodland Caribou
    (*Rangifer tarandus caribou*), Boreal Population, in Canada." Species at
    Risk Act: Recovery Strategy Series, 2012. http://www.sararegistry.gc.ca
    /document/default_e.cfm?documentID=2253.
Evers, David C. *Conserve the Call: Identifying and Managing Environmental
    Threats to the Common Loon.* Portland, ME: Biodiversity Research Institute's
    Center for Loon Conservation. Science Communications Series BRI
    2014-21, 2014.
Evers, David C., Joseph D. Kaplan, Michael W. Meyer, Peter S. Reaman,
    W. Emmett Braselton, Andrew Major, Neil Burgess, and Anton M.
    Scheuhammer. "Geographic Trend in Mercury Measured in Common
    Loon Feathers and Blood." *Environmental Toxicology and Chemistry* 17, no. 2
    (1998): 173–83.
Fashingbauer, Bernard. "The Woodland Caribou in Minnesota." In *Big Game
    in Minnesota,* edited by J.B. Moyle, 133–66. Minnesota Department of
    Conservation Technical Bulletin, no. 9, 1965.
Flader, Susan L. *Thinking Like a Mountain: Aldo Leopold and the Evolution of an
    Ecological Attitude Toward Deer, Wolves, and Forests.* Columbia: University of
    Missouri Press, 1975.
Fletcher, Tom. "B.C. Interior Caribou Protection Area Big Enough, Minister
    Says." *Victoria News,* September 18, 2019. https://www.vicnews.com/news
    /b-c-interior-caribou-protection-area-big-enough-minister-says/.
Foster, J. W. "Recent Advances in Geology." *The American Naturalist* 4, no. 8
    (1870): 449–72.
Francovich, Eli, "South Selkirk Mountain Caribou Herd Possibly Extinct,"

Spokesman-Review, April 21, 2018; accessed August 8, 2020. https://www
.spokesman.com/stories/2018/apr/19/south-selkirk-mountain-caribou-herd
-possibly-extin/.

Frechette, James, and Mike Hoffman. "The Menominee Clans Story."
University of Wisconsin-Stevens Point. Accessed July 23, 2020. https://
www4.uwsp.edu/museum/menomineeClans/origin/.

Friedman, Amy, and Meredith Johnson. "The Gift of the Loon: An Inuit
Legend." *UExpress*, August 22, 2010. http://www.uexpress.com/tell-me-a
-story/2010/8/22/the-gift-of-the-loon-an.

Fuller, P. "Lake Sturgeon (*Acipenser fulvescens*)—Species Profile." U.S.
Geological Survey, July 1, 2019. https://nas.er.usgs.gov/queries/FactSheet.
aspx?SpeciesID=299.

Gardiner, Brian G. "Sturgeons as Living Fossils." In *Living Fossils*, edited by
N. Eldredge and S.M. Stanley, 148–52. New York: Springer, 1984. https://doi
.org/10.1007/978-1-4613-8271-3_15.

Geist, Valerius. "Of Reindeer and Man, Modern and Neanderthal: A Creation
Story Founded on a Historic Perspective on How to Conserve Wildlife,
Woodland Caribou in Particular." *Rangifer*, April 1, 2003, 57–63. https://doi
.org/10.7557/2.23.5.1681.

Gelernter, David. "Does the Universe Have A Purpose?" The Templeton
Foundation, n.d. https://www.ethics-based-on-science.com/uploads
/2/8/5/1/28516163/mjm-notes-does-the-universe-have-a-purpose-bq
_universe-1.pdf.

Gogan, Peter J. P., and Jean Fitts Cochrane. "Restoration of Woodland
Caribou to the Lake Superior Region." U.S. National Park Service
Publications and Papers, January 1, 1994. http://digitalcommons.unl.edu
/natlpark/11.

Grayson, Donald, and Francoise Delpech. "Pleistocene Reindeer and Global
Warming." *Conservation Biology* 19, no. 2 (2005): 6.

Gross, Liza. "Feeding Wild Birds Can Carry Risks: Here's How to Minimize
Unintended Harms." *PBS* (blog), September 10, 2019. https://www.pbs.org
/wnet/nature/blog/feeding-wild-birds-can-carry-risks-heres-how-to-
minimize-unintended-harms/.

Grumbine, R. Edward. *Ghost Bears: Exploring The Biodiversity Crisis.*
Washington, D.C: Island Press, 1992.

Guynup, Sharon. "Loons Sound Alarm on Mercury Contamination." *National
Geographic Today*, May 16, 2013.

Gwich'in Steering Committee. "A Moral Choice for the United States: The

Human Rights Implications for the Gwich'in of Drilling in the Arctic National Wildlife Refuge." Fairbanks, AK, 2005. https://episcopalchurch .org/files/documents/gschumanrightsreport.pdf.

Hannibal-Paci, Christopher J. "His Knowledge and My Knowledge: Cree and Ojibwe Traditional Environmental Knowledge and Sturgeon Co-Management in Manitoba," Ph.D. dissertation, University of Manitoba, July 1, 2000. https://mspace.lib.umanitoba.ca/xmlui/handle/1993/1867.

———. "Historical Representations of Lake Sturgeon by Native and Non-Native Artists." *Canadian Journal of Native Studies* 18, no. 2 (1998): 214.

Hanson, Jeanne K. "The Extraordinary Life of Étienne Brûlé: Breaking Trail to the Big Lake in the 17th Century." *Lake Superior Magazine*, June 1, 2014. https://www.lakesuperior.com/the-lake/great-lakes/etienne-brule-breaking -trail-to-the-big-lake-in-the-17th-century/.

Harding, Lee E., Mathieu Bourbonnais, Andrew T. Cook, Toby Spribille, Viktoria Wagner, and Chris Darimont. "No Statistical Support for Wolf Control and Maternal Penning as Conservation Measures for Endangered Mountain Caribou." *Biodiversity and Conservation*, July 14, 2020. https:// doi.org/10.1007/s10531-020-02008-3.

Hardy, Campbell. *Forest Life in Acadie : Sketches of Sport and Natural History in the Lower Provinces of the Canadian Dominion.* London: Chapman & Hall, 1869. http://archive.org/details/forestlifeinacaoohard.

Harland-Haughey, Sarah. "The Broken Bird: Note on the Unsolved Mystery of the Loon's Name." *Spire, the Maine Journal of Conservation and Sustainability* 1 (2017). https://umaine.edu/spire/2017/05/04/harlan -haughey/.

Hassell, Karli Tyance. "Two-Eyed Seeing: Science, Indigenous Knowledge, and Partnership in the English Bay Lakes System, Alaska." Ph.D. thesis, Alaska Pacific University, 2019.

Heasley, Lynne. *The Accidental Reef and other Ecological Odysseys in the Great Lakes.* East Lansing, MI: Michigan State University Press, 2021.

Henry, Alexander. *Alexander Henry's Travels and Adventures in the Years 1760–1776.* Chicago: Lakeside Press, 1921.

Hobgood-Oster, Laura. *A Dog's History of the World: Canines and the Domestication of Humans.* Waco, TX: Baylor University Press, 2014.

Holtgren, Marty, Stephanie Ogren, and Kyle Whyte. "Renewing Relatives." *Earth Island Journal*, 2015. http://www.earthisland.org/journal/index.php /magazine/entry/renewing_relatives/.

Holzkamm, Tim E. "Sturgeon Utilization by the Rainy River Ojibwa Bands."

Papers of the Algonquian Conference 18 (1987):155–63. https://ojs.library
.carleton.ca/index.php/ALGQP/article/view/946/830.

Holzkamm, Tim E., Victor P. Lytwyn, and Leo G. Waisberg. "Rainy River
Sturgeon: An Ojibway Resource in the Fur Trade Economy." *The Canadian
Geographer* 32, no. 3 (September 1, 1988): 194–205. https://doi.org
/10.1111/j.1541-0064.1988.tb00873.x.

Holzkamm, Tim E., and Leo G. Waisberg. "Native American Utilization of
Sturgeon." In *Sturgeons and Paddlefish of North America* edited by G.T.O
LeBreton, F.W.H. Bemaish, and R.S. McKinley, 22–39. New York: Kluwer
Academic Publishers, 2004. https://doi.org/10.1007/1-4020-2833-4_2.

Holzkamm, Tim, and Michael McCarthy. "Potential Fishery for Lake
Sturgeon (*Acipenser fulvescens*) as Indicated by the Returns of the Hudson's
Bay Company Lac La Pluie District." *Canadian Journal of Fisheries and
Aquatic Sciences* 45, no. 5 (1998): 921–23. https://doi.org/10.1139/f88-113.

Huff, A., and A. Thomas, "Lake Superior Climate Change Impacts and
Adaptation." Prepared for the Lake Superior Lakewide Action and
Management Plan—Superior Work Group, 2014. https://www.epa.gov
/greatlakes/lake-superior-climate-change-impacts-report.

Imbler, Sabrina. "How a Simple Statistical Error Killed 463 Wolves." *The
Atlantic*, July 14, 2020. https://www.theatlantic.com/science/archive
/2020/07/how-simple-statistical-error-killed-463-wolves/614134/.

Jacoby, Karl. *Crimes Against Nature: Squatters, Poachers, Thieves, and the Hidden
History of American Conservation.* Berkeley: University of California Press, 2001.

Johnsgard, Paul A. *The Niobrara: A River Running Through Time.* Lincoln:
University of Nebraska Press, 2007.

Johnson, Charles Eugene. "Recollections of the Mammals of Northwestern
Minnesota." *Journal of Mammalogy* 11, no. 4 (1930): 435–52.

Jorgensen, Dolly. "After None: Memorialising Animal Species Extinction
through Monuments." In *Animals Count: How Population Size Matters in
Animal-Human Relations*, edited by Nancy Cushing and Jodi Frawley, 183–
99. London: Routledge, 2018. https://dolly.jorgensenweb.net/after-none
-memorialising-animal-species-extinction-through-monuments/.

Kallok, Michael. "A Whopper of a Recovery." *Minnesota Conservation
Volunteer*, 2017. https://www.dnr.state.mn.us/mcvmagazine/issues/2017/may
-jun/lake-sturgeon-restoration.html.

Kimmerer, Robin. *Braiding Sweetgrass: Indigenous Wisdom, Scientific Knowledge
and the Teachings.* Minneapolis: Milkweed Editions, 2013.

King, J.C.H., Birgit Pauksztat, and Robert Storrie. *Arctic Clothing*. Montreal: McGill-Queen's University Press, 2005.

Kingsley, Patrick, and David Levene. "Nomads No More: Why Mongolian Herders Are Moving to the City." *The Guardian*, January 5, 2017. https:// www.theguardian.com/world/2017/jan/05/mongolian-herders-moving-to -city-climate-change.

Klein, Tom. "Loonacy." *Chicago Tribune*, March 23, 1986. https://www .chicagotribune.com/news/ct-xpm-1986-03-23-8601210583-story.html.

Kline, Kathleen Schmitt. *People of the Sturgeon: Wisconsin's Love Affair with an Ancient Fish*. Madison: University of Wisconsin Press, 2009.

Koontz, Tomas M., and Jennifer Bodine. "Implementing Ecosystem Management in Public Agencies: Lessons from the U.S. Bureau of Land Management and the Forest Service." *Conservation Biology* 22, no. 1 (February 1, 2008): 60–69. https://doi.org/10.1111/j.1523-1739.2007.00860.x.

Kuehn, Steven R. "New Evidence for Late Paleoindian-Early Archaic Subsistence Behavior in the Western Great Lakes." *American Antiquity* 63, no. 3 (1998): 457–76.

"Lake Sturgeon Rehabilitation," Gun Lake Tribe, 2017. https://gunlaketribe -nsn.gov/departments/administration/environmental/lake-sturgeon -rehabilitation/.

"Lake Sturgeon Restoration." *West Michigan Conservation Network* (blog), March 6, 2020. https://wmconservation.net/2020/03/06/lake-sturgeon -restoration/.

Langston, Nancy. "Are Woodland Caribou Doomed by Climate Change?" *Historical Climatology*, July 26, 2018. http://www.historicalclimatology.com /1/post/2018/07/are-woodland-caribou-doomed-by-climate-change.html.

———. "Environmental History and Restoration in the Western Forests." *Journal of the West* 38, no. 4 (1999): 45–56.

———. *Forest Dreams, Forest Nightmares: The Paradox of Old Growth in the Inland West*. Seattle: University of Washington Press, 1995.

———. "Mining the Boreal North." *American Scientist*, June 2016. https:// www.americanscientist.org/article/mining-the-boreal-north.

———. "Paradise Lost: Climate Change, Boreal Forests, and Environmental History." *Environmental History* 14, no. 4 (October 1, 2009): 641–50. https:// doi.org/10.1093/envhis/14.4.641.

———. *Sustaining Lake Superior: An Extraordinary Lake in a Changing World*. New Haven: Yale University Press, 2017.

————. *Toxic Bodies: Hormone Disruptors and the Legacy of DES*. New Haven: Yale University Press, 2010.

————. *Where Land and Water Meet: A Western Landscape Transformed*. Seattle: University of Washington Press, 2003.

————. "Will Woodland Caribou Survive in the Lake Superior Basin?" *Agate*, January 14, 2019. http://www.agatemag.com/2019/01/will-woodland -caribou-survive-in-the-lake-superior-basin/.

Langston, Nancy, and Kate Christen. "Conservation Policies Threaten Indigenous Reindeer Herders in Mongolia." *The Conversation*, 2019. http:// theconversation.com/conservation-policies-threaten-indigenous-reindeer -herders-in-mongolia-121729.

Latour, Bruno. *We Have Never Been Modern*. Cambridge, MA: Harvard University Press, 2012.

Laundre, John W., Lucina Hernandez, and William J. Ripple. "The Landscape of Fear: Ecological Implications of Being Afraid." *The Open Ecology Journal* 3, no. 1 (February 3, 2010). https://benthamopen.com/ABSTRACT/TOE COLJ-3-3-1.

Leopold, Aldo. *A Sand County Almanac and Sketches Here and There*. New York: Oxford University Press, 1949.

Lescureux, Nicolas. "Beyond Wild and Domestic: Human Complex Relationships with Dogs, Wolves, and Wolf-Dog Hybrids." In *Hybrid Communities: Biosocial Approaches to Domestication and Other Trans-Species Relationships*, edited by Charles St´panoff and Jean-Denis Vigne. London: Routledge, 2020.

Levinovitz, Alan. *Natural: How Faith in Nature's Goodness Leads to Harmful Fads, Unjust Laws, and Flawed Science*. Boston: Beacon Press, 2020.

Lewis, Daniel. *Belonging on an Island: Birds, Extinction, and Evolution in Hawai'i*. New Haven: Yale University Press, 2018.

Lewis, John. "Deer to Our Culture." *Wisconsin Natural Resources Magazine*, December 1998.

Loew, Patty. *Indian Nations of Wisconsin: Histories of Endurance and Renewal*. Madison: Wisconsin Historical Society Press, 2013.

Loew, Patty, and James Thannum. "After the Storm: Ojibwe Treaty Rights Twenty-Five Years after the Voigt Decision." *The American Indian Quarterly* 35, no. 2 (2011): 161–91.

Lopez, Barry Holstun. *Of Wolves and Men*. New York: Charles Scribner's Sons, 1979.

Losey, Robert J., Tatiana Nomokonova, Lacey Fleming, Katherine Latham, and Lesley Harrington. "Domestication and the Embodied Human-Dog Relationship: Archaeological Perspectives from Siberia." In *Dogs in the North: Stories of Cooperation and Co-Domestication*, edited by Robert J. Losey, Robert P. Wishart, and Jan Peterlaurens Loovers. London: Routledge, 2018. https://doi.org/10.4324/9781315437736-2.

Maher, Neil M. *Nature's New Deal: The Civilian Conservation Corps and the Roots of the American Environmental Movement*. New York: Oxford University Press, 2009.

Manweiler, John. "Minnesota's Woodland Caribou." *The Conservation Volunteer* 1, no. 4 (1941): 34–40.

———. "Wildlife Management in Minnesota's 'Big Bog.'" *The Minnesota Conservationist*, 1938, 14–15.

———. "Woodland Caribou from Saskatchewan." *Parks and Recreation* 22, no. 3 (1938): 134–38.

———. "Woodland Caribou in the Big Bog." *The Minnesota Conservationist* 65 (1939): 16–17, 23, 30.

———. "Woodland Caribou Study in Northern Minnesota." *Parks and Recreation* 22, no. 2 (1938): 74–78.

McCance, Dawne. *Critical Animal Studies: An Introduction*. Albany: State University of New York Press, 2013.

McCann, Patricia. "Mercury Levels in Blood from Newborns in Lake Superior Basin." Minnesota Department of Health (MDH) Fish Consumption Advisory Program and MDH Public Health Laboratory, GLPNO ID 2007–942, November 30, 2011.

McClellan, Catharine. *My Old People Say: An Ethnographic Survey of Southern Yukon Territory*. National Museums of Canada, 1975.

McCormick, Damon L., Joseph D. Kaplan, and Keren Tischler. "Common Loon Research at Seney National Wildlife Refuge: 2007 Field Season." Common Coast Research Conservation, 2007. https://www.fws.gov/uploadedFiles/McCormick2007.pdf.

———. "Mercury Exposure in Common Loons at Seney NWR." Common Coast Research Conservation, January 26, 2006. https://catalog.data.gov/dataset/mercury-exposure-in-common-loons-at-seney-nwr.

McEvoy, Arthur F. *The Fisherman's Problem: Ecology and Law in the California Fisheries, 1850–1980*. Cambridge: Cambridge University Press, 1986.

McInnes, Brian D. *Sounding Thunder: The Stories of Francis Pegahmagabow*. Winnipeg: University of Manitoba Press, 2016.

McMillan, L. Jane, and Kerry Prosper. "Remobilizing Netukulimk: Indigenous Cultural and Spiritual Connections with Resource Stewardship and Fisheries Management in Atlantic Canada." *Reviews in Fish Biology and Fisheries* 26, no. 4 (2016): 629–47.

McNeill, J. R., and Peter Engelke. *The Great Acceleration: An Environmental History of the Anthropocene since 1945*. Cambridge, MA: Harvard University Press, 2016.

Mele, Andre, and Audre Mele. *Polluting for Pleasure*. New York: W. W. Norton & Co., 1993.

Meyer, Melissa L. "The Red Lake Ojibwe." In *Patterned Peatlands of Minnesota*, edited by Herbert Edgar Wright, Barbara Coffin, and Norman Aaseng, 251–62. Minneapolis: University of Minnesota Press, 1992.

Milner. "U.S. Fish Commission: Report of the Commissioner for 1872–73," 1874. http://penbay.org/cof/cof_1872_1873.html.

Murray, William. *Adventures in the Wilderness or Camp-Life in the Adirondacks*. Boston: Fields, Osgood & Co., 1869.

Nelson, Richard. "Eskimo Science." In *Conformity and Conflict: Readings in Cultural Anthropology*, edited by James P. Spradley and David W. McCurdy. Boston: Pearson, 2012.

———. *Make Prayers to the Raven: A Koyukon View of the Northern Forest*. Chicago: University of Chicago Press, 2020.

Nickens, T. Edward. "Paper Chase." *Audubon Magazine*, January-February 2009. https://www.audubon.org/magazine/january-february-2009/paper-chase.

Nootchtai, M. Why the Loon Can't Walk. In E. Higgins, *Nookomis O Dibajamonwin (Grandmother Tell Me a Story)*. Cobalt, Ontario: Highway Books, 1986.

Olson, Sigurd F. *Singing Wilderness*. New York: Knopf, 1956.

Olson, Storrs L., Horace Loftin, and Steve Goodwin. "Biological, Geographical, and Cultural Origins of the Loon Hunting Tradition in Carteret County, North Carolina." *The Wilson Journal of Ornithology* 122, no. 4 (December 1, 2010): 716–24. https://doi.org/10.1676/10-048.1.

Packer, Anthony. "Manitoba History: Glacial Lake Agassiz." *Manitoba History* 19(Spring 1990). http://www.mhs.mb.ca/docs/mb_history/19/lakeagassiz.shtml.

Padilla, Elisabeth. "Caribou Leadership: A Study of Traditional Knowledge, Animal Behavior, and Policy." Ph.D. thesis, University of Alaska Fairbanks, 2010.

Padilla, Elisabeth, and Gary P. Kofinas. "'Letting the Leaders Pass': Barriers to Using Traditional Ecological Knowledge in Comanagement as the Basis of Formal Hunting Regulations." *Ecology and Society* 19, no. 2 (2014).

Page, Julia. "Efforts to Save Woodland Caribou in Northern Quebec Too Costly, Says Province." *CBC News*, March 9, 2018. https://www.cbc.ca /news/canada/montreal/efforts-to-save-woodland-caribou-in-northern -quebec-too-costly-says-province-1.4569479.

Palmer, Claire. "Profiles." *Environmental Humanities* (blog), 2019. http:// environmentalhumanities.org/about/profiles.

Parker, John, ed. *The Journals of Jonathan Carver and Related Documents, 1766– 1770*. St. Paul: Minnesota Historical Society Press, 1976. http://archive.org /details/journalsofjonathoocarv.

Peers, Laura. "Ontario Paleo-Indians and Caribou Predation." *Ontario Archaeology* 43 (1985): 31–40.

Piper, Liza, and John Sandlos. "A Broken Frontier: Ecological Imperialism in the Canadian North." *Environmental History* 12, no. 4 (2007): 759–795.

"Plan to Solve Great Lakes Problem." *Winnipeg Evening Tribune*, Friday November 6, 1925, p. 14.

Prince, Hugh. *Wetlands of the American Midwest: A Historical Geography of Changing Attitudes*. Chicago: University of Chicago Press, 2008.

Pritchard, Sara B. "Joining Environmental History with Science and Technology Studies." In *New Natures: Joining Environmental History with Science and Technology Studies*, edited by Dolly Jørgensen, Finn Arne Jørgensen, and Sara B. Pritchard, 1–20. Pittsburgh, PA: University of Pittsburgh Press, 2013.

Northland College. "Protect Loons." Accessed August 11, 2020. https://www .northland.edu/centers/soei/loonwatch/protect-loons/.

Michigan Sea Grant. "Avian Botulism." Accessed January 5, 2021. https://www .michiganseagrant.org/topics/coastal-hazards-and-safety/avian-botulism/.

Minnesota Department of Natural Resources. "Red Lake WMA." Accessed July 22, 2020. https://www.dnr.state.mn.us/areas/wildlife/red_lake_wma .html.

Pennisi, Elizabeth. "Three Billion North American Birds Have Vanished since 1970, Surveys Show." *Science*, September 19, 2019. https://www.sciencemag

.org/news/2019/09/three-billion-north-american-birds-have-vanished
-1970-surveys-show.

Rayne, Aisling, Greg Byrnes, Levi Collier-Robinson, John Hollows, Angus McIntosh, Mananui Ramsden, Makarini Rupene, Paulette Tamati-Elliffe, Channell Thoms, and Tammy E. Steeves. "Centring Indigenous Knowledge Systems to Re-Imagine Conservation Translocations." *People and Nature* 2, no. 3 (2020): 512–26.

Rees, Amanda. "Can Animals Shape Their Own Lives? Or the Course of History?" *Aeon*, February 26, 2018. https://aeon.co/essays/can-animals -shape-their-own-lives-or-the-course-of-history.

Reid, Andrea J., Lauren E. Eckert, John-Francis Lane, Nathan Young, Scott G. Hinch, Chris T. Darimont, Steven J. Cooke, Natalie C. Ban, and Albert Marshall. "'Two-Eyed Seeing': An Indigenous Framework to Transform Fisheries Research and Management." *Fish and Fisheries*, 2020.

Reilly, Amanda. "EPA 'Still Thinking About' Obama Mercury Standards— Wehrum." *E&E News*, April 19, 2018. https://www.eenews.net/stories /1060079569.

"Resilience of the Lake Sturgeon." Accessed July 23, 2020. https://www.arcgis .com/apps/MapJournal/index.html?appid=a7644a96ab00491e9cfa070 b28b60467&=&utm_source=SocialMedia&utm_medium=Social Media&utm_campaign=resilience.

Riis, Paul. "Woodland Caribou and Time: Part 1." *Parks and Recreation* 21 (1938): 521, 529–35; 594–600; 639–45.

Saskatchewan Environmental Society. "Climate Change and Saskatchewan's Boreal Forest: A Saskatchewan Environmental Society Fact Sheet." December 2003.

Roosevelt, Theodore. *The Deer Family*. New York: Macmillan and Co., 1902. http://hdl.handle.net/2027/loc.ark:/13960/t2w38d11c.

Rosenfield, Leonora Davidson. *From Beast-Machine to Man-Machine: Animal Soul in French Letters from Descartes to La Mettrie*. London, U.K.: Octagon Books, 1968.

Rosentreter, Roger. "Roosevelt's Tree Army: Michigan's Civilian Conservation Corps." *Michigan History Magazine* (May/June 1986). Republished online by the Michigan History Center. https://www.michigan.gov/mhc/0,9075 ,7-361-85147_87219_87222-472998--,00.html.

Rosner, Hillary. "Pulling Canada's Caribou Back from the Brink." *The Atlantic*, December 17, 2018. https://www.theatlantic.com/science/archive/2018/12 /heroic-measures-for-canadas-caribou/577789/.

Rudolf, Paul O. *History of the Lake States Forest Experiment Station*. North
    Central Forest Experiment Station, Forest Service, U.S. Department of
    Agriculture, 1985.

Rulseh, Ted. "Happy Anniversary, LoonWatch." *Lakeland Times*, May 8, 2018.
    https://www.northland.edu/news/happy-anniversary-loonwatch/.

Samuels, Edward A. *With Rod and Gun in New England and the Maritime
    Provinces*. Boston: Samuels & Kimball, 1897. https://catalog.hathitrust.org
    /Record/002025623.

Schoolcraft, Henry Rowe. *Historical and Statistical Information Respecting
    the History, Condition and Prospects of the Indian Tribes of the United States*.
    Philadelphia: Lippincott, Grambo & Co., 1851. https://doi.org/10.5479
    /sil.131145.39088002742823.

Schulz, Kathryn. "The Earthquake That Will Devastate the Pacific
    Northwest." *The New Yorker*, July 13, 2015. https://www.newyorker.com
    /magazine/2015/07/20/the-really-big-one.

Scott, James C. *Seeing Like a State: How Certain Schemes to Improve the Human
    Condition Have Failed*. New Haven: Yale University Press, 1998.

"Seney National Wildlife Refuge Annual Narrative Report [1965]." Accessed
    October 22, 2019. https://catalog.data.gov/dataset/seney-national-wildlife
    -refuge-annual-narrative-report-1965.

Seno, William Joseph, ed. *Up Country: Voices from the Midwestern Wilderness*.
    Madison, WI: Round River Press, 1985.

Serrouya, Robert, M. Dickie, C. DeMars, M.J. Wittmann, and S. Boutin.
    "Predicting the Effects of Restoring Linear Features on Woodland Caribou
    Populations." *Ecological Modelling* 416 (January 2020): 108891. https://doi
    .org/10.1016/j.ecolmodel.2019.108891.

Serrouya, Robert, Dale R. Seip, Dave Hervieux, Bruce N. McLellan,
    R. Scott McNay, Robin Steenweg, Doug C. Heard, Mark Hebblewhite,
    Michael Gillingham, and Stan Boutin. "Saving Endangered Species Using
    Adaptive Management." *Proceedings of the National Academy of Sciences* 116,
    no. 13 (March 26, 2019): 6181–86. https://doi.org/10.1073/pnas.1816923116.

Sheehan, John F. "Acid Rain, Mercury and Loons . . . Oh My!" Adirondack
    Council Blog, June 30, 2014. https://www.adirondackcouncil.org/page/blog
    -139/news/acid-rain-mercury-and-loonsoh-my-490.html.

Shields, G. O., ed. *The Big Game of North America. Its Habits, Habitats,
    Haunts, and Characteristics; How, When, and Where to Hunt It*. Chicago:
    Rand, McNally & Co., 1890. https://catalog.hathitrust.org
    /Record/001509719.

Shipman, Pat. *The Animal Connection: A New Perspective On What Makes Us Human*. New York: W.W. Norton, 2011.

Siewert, Shereen. "In Wisconsin, Hate Makes a Comeback." *Wausau Pilot & Review*, June 15, 2020. https://wausaupilotandreview.com/2020/06/15 /in-wisconsin-hate-makes-a-comeback/.

Simkin, Daniel. "Fish and Wildlife Management Report, April 1, 1960." Ontario Department of Lands and Forests. http://archive.org/details /resourcemanapr1960onta.

Smith, Mick. "Profiles." *Environmental Humanities* (blog). Accessed March 8, 2019. http://environmentalhumanities.org/about/profiles/ep-smith.

Solnoi, Batulag, Purev Tsogtsalkhan, and Dan Plumley. "Following the White Stag: The Dukha and Their Struggle for Survival." *Cultural Survival Quarterly* 27, no. 1 (2003). https://www.culturalsurvival.org/publications /cultural-survival-quarterly/following-white-stagthe-dukha-and-their -struggle-survival.

Sonnenburg, Elizabeth, Ashley K. Lemke, and John M. O'Shea, eds. *Caribou Hunting in the Upper Great Lakes: Archaeological, Ethnographic, and Paleoenvironmental Perspectives*. Memoirs of the Museum of Anthropology, University of Michigan, no. 57. Ann Arbor: The Museum of Anthropology, University of Michigan, 2015.

Sorenson, John. *Critical Animal Studies: Thinking the Unthinkable*. Toronto: Canadian Scholars' Press Inc, 2014.

Spence, Mark David. *Dispossessing the Wilderness: Indian Removal and the Making of the National Parks*. New York: Oxford University Press, 1999.

Spratt, Ashley. "The Return of Namé 'King of Fish' to the Red River Basin." U.S. Fish and Wildlife Service Midwest Region, 2017. https://www.fws.gov /midwest/kingoffish.htm.

Steffen, Will, Paul J. Crutzen, and John R. McNeill. "The Anthropocene: Are Humans Now Overwhelming the Great Forces of Nature." *AMBIO: A Journal of the Human Environment* 36, no. 8 (December 2007): 614–22. https://doi.org/10.1579/0044-7447(2007)36[614:TAAHNO]2.0.CO;2.

Steinke, David. *Green Fire: Aldo Leopold and a Land Ethic for Our Time*, 2011. https://www.aldoleopold.org/teach-learn/green-fire-film/.

Storck, Peter L., and Arthur E. Spiess. "The Significance of New Faunal Identifications Attributed to an Early Paleoindian (Gainey Complex) Occupation at the Udora Site, Ontario, Canada." *American Antiquity* 59, no. 1 (January 1994): 121–42. https://doi.org/10.2307/3085506.

Sulak, K. J., R. E. Edwards, G. W. Hill, and M. T. Randall. "Why Do Sturgeons Jump? Insights from Acoustic Investigations of the Gulf Sturgeon in the Suwannee River, Florida, USA." *Journal of Applied Ichthyology* 18, nos. 4–6 (2002): 617–20. https://doi:10.1046/j.1439-0426.2002.00401.x.

"Survival by Degrees: 389 Bird Species on the Brink." National Audubon Society. Accessed December 5, 2019. https://www.audubon.org/climate/survivalbydegrees.

Swanson, Gustav. "The Minnesota Caribou Herd." *Proceedings of the North American Wildlife Conference Called by President Franklin D. Roosevelt*, 1936, 415–18.

Tanner, John. *The Falcon: A Narrative of the Captivity and Adventures of John Tanner.* New York: Penguin Books, 1994. http://archive.org/details/falconnarrativeoooootann.

Taylor, Joseph E. *Making Salmon: An Environmental History of the Northwest Fisheries Crisis.* Seattle: University of Washington Press, 2001.

Te Beest, Mariska, Judith Sitters, Cécile B. Ménard, and Johan Olofsson. "Reindeer Grazing Increases Summer Albedo by Reducing Shrub Abundance in Arctic Tundra." *Environmental Research Letters* 11, no. 12 (2016): 125013.

Thoreau, Henry David. *Walden; Or, Life in the Woods.* New York: Ticknor and Fields, 1854. Reprint, New York: Dover Publications, 2012.

Thorpe, J., N. Henderson, and J. Vandall. *Ecological and Policy Implications of Introducing Exotic Trees for Adaptation to Climate Change in the Western Boreal Forest.* Saskatchewan Research Council Publication 11776-1E06, Saskatoon, Saskatchewan, 2006).

Tozer, Douglas C., C. Myles Falconer, and Debbie S. Badzinski. "Common Loon Reproductive Success in Canada: The West Is Best but Not for Long." *Avian Conservation and Ecology* 8, no. 1 (2013). https://doi.org/10.5751/ACE-00569-080101.

Treuer, Anton. *Warrior Nation: A History of the Red Lake Ojibwe.* St. Paul: Minnesota Historical Society Press, 2015.

University of Wisconsin-Madison Arboretum. "Caribou Migration." Journey North: Tracking Migrations and Seasons. Accessed July 22, 2020. https://journeynorth.org/tm/caribou/BuildACaribou.html.

U.S. Geological Survey, Upper Midwest Environmental Sciences Center. "Common Loon Migration Study—Update," April 3, 2018. https://www

.umesc.usgs.gov/terrestrial/migratory_birds/loons/update.html.

———. "Loon Study—Frequently Asked Questions." Accessed July 4, 2020. https://www.umesc.usgs.gov/terrestrial/migratory_birds/loons/questions .html.

U.S. Government. "Treaty with the Ottawa, Etc. Articles of a treaty made and concluded at the city of Washington in the District of Columbia, between Henry R. Schoolcraft, commissioner on the part of the United States, and the Ottawa and Chippewa nations of Indians, by their chiefs and delegates." March 28, 1836. Available at http://www.1836cora.org/

Vitebsky, Piers. *The Reindeer People: Living with Animals and Spirits in Siberia.* New York: Houghton Mifflin Harcourt, 2006.

Vucetich, John, and Michael Nelson. "Wolf Hunting and the Ethics of Predator Control." In *The Oxford Handbook of Animal Studies*, edited by Linda Kalof. New York: Oxford University Press, 2017.

Waal, Frans De. *The Ape and the Sushi Master.* New York: Basic Books, 2001.

Waldman, John. "The Lofty Mystery of Why Sturgeon Leap." *The New York Times*, October 21, 2001. https://www.nytimes.com/2001/10/21/sports /outdoors-the-lofty-mystery-of-why-sturgeon-leap.html.

Weber, Bob. "Increase Wolf Cull, Pen Pregnant Cows to Save Endangered Caribou." *CBC News*, March 11, 2019. https://www.cbc.ca/news/canada /edmonton/increase-wolf-cull-to-save-endangered-caribou-1.5051901.

Weil, Kari. *Thinking Animals: Why Animal Studies Now?* New York: Columbia University Press, 2012.

Westley, Peter A. H., Andrew M. Berdahl, Colin J. Torney, and Dora Biro. "Collective Movement in Ecology: From Emerging Technologies to Conservation and Management." *Philosophical Transactions of the Royal Society B: Biological Sciences* 373, no. 1746 (May 19, 2018): 20170004. https:// doi.org/10.1098/rstb.2017.0004.

Westropp, Hodder M. "On the Sequence of the Phases of Civilisation, and Contemporaneous Implements." *Journal of the Anthropological Society of London* 5 (1867): cxcii–cc. https://doi.org/10.2307/3025265.

White, Richard. *The Middle Ground: Indians, Empires, and Republics in the Great Lakes Region, 1650–1815.* Cambridge: Cambridge University Press, 1991.

———. *The Organic Machine: The Remaking of the Columbia River.* New York: Hill & Wang, 1995.

Whitney, Caspar. *On Snow-Shoes to the Barren Grounds; Twenty-Eight Hundred Miles after Musk-Oxen and Wood-Bison.* New York: Harper & Brothers, 1896. https://catalog.hathitrust.org/Record/000204580.

Williams, Ted. "Recovery: Saving Lake Sturgeon, an Ancient Fish with a Bright Future." *Cool Green Science*, February 15, 2016. https://blog.nature .org/science/2016/02/15/recovery-saving-lake-sturgeon-ancient-fish-bright -future/.

Willis, Roxanne. "A New Game in the North: Alaska Native Reindeer Herding, 1890–1940." *Western Historical Quarterly* 37, no. 3 (2006): 277–301. https://doi.org/10.2307/25443371.

Wills, Matthew. "The Reindeer Games." *JSTOR Daily*, December 9, 2019. https://daily.jstor.org/the-reindeer-games/.

Wilson, Robert M. "Mobile Bodies: Animal Migration in North American History." *Geoforum* 65 (October 1, 2015): 465–72. https://doi.org/10.1016/j .geoforum.2015.04.001.

Wray, Kristine, and Brenda Parlee. "Ways We Respect Caribou: Teetł'it Gwich'in Rules." *Arctic*, 2013, 68–78.

Yaktine, Ann L., Gail G. Harrison, and Robert S. Lawrence. "Reducing Exposure to Dioxins and Related Compounds through Foods in the Next Generation." *Nutrition Reviews* 64, no. 9 (September 2006): 403–9. https:// doi.org/10.1111/j.1753-4887.2006.tb00225.x.

INDEX

Note: Images in text are indicated here in *italics*.

Treaty of 1863, 30
treaty rights, 30, 75, 86–88, 96–97, 116, 147n30
tribal biologists, 98–99
tribal hunting, 37–38
tribal identities, 81
tribal lands, 30–34, 36–37
tribal restoration programs, 96–100
Trump administration, 126, 128
two-eyed seeing, 106

Ulaanbaatar, Mongolia, 65–66
U.S. Supreme Court, 128

Val-d'Or caribou, 53
vaulting, 79
Vitebsky, Piers, 59, 65
vocalizations, 118
Voigt Decision, 96
Volstead Act, 32

Walden (Thoreau), 114
Waldman, John, 75
wallows, 34
Warren, W. W., 24
Washington Treaty of 1836, 116
Wawa, Ontario, 12, 51
weather, 13
Western Lake Superior Sanitary District, 100
Westropp, Hodder M., 61
wetlands environment, 33

White, Richard, 93–94
White Earth Band of Ojibwe, 31, 81–82, 99–100
White Earth Reservation, 31
White Horse, 93
white-tailed deer (*Odocoileus virginianus*), 17–18, 20–21, 30, 56–57
Whitney, Caspar, 26
Why Look at Animals (Berger), 60
wilderness, 68, 72, 109, 145n1
wildlife management, 41
wildness, 104, 114
wild rice or *manoomin* (*Zizania palustris*), 81
Wilson, Robert, 2–3
winter precipitation, 56–57
Wisconsin, 75, 88, 91, 96
Wisconsin Department of Natural Resources, 102
Wolf River, 101–3
wolves: and culls, 49–54; and death, 113–14; and drainage ditches, 40–41; and Euro-Americans, 51, 133n11; and predation of woodland caribou, 16–17, 19–21, 36–38, 44–45, 46–47, 49–54; and the Slate Islands, 13
Wood, Warren, 31–32

Yellowstone, 49
Yosemite National Park, 66
Yukon, 21